Engineering Money

Engineering Money
Financial Fundamentals for Engineers

Richard Hill
George Solt

A John Wiley & Sons, Inc., Publication

For general information on our other products and services or for technical support, please
contact our Customer Care Department within the United States at (800) 762-2974, outside the
United States at (317) 572-3993 or fax (317) 572-4002.

Wiley also publishes its books in a variety of electronic formats. Some content that appears in
print may not be available in electronic formats. For more information about Wiley products,
visit our web site at www.wiley.com.

Library of Congress Cataloging-in-Publication Data:

Hill, Richard (Richard William), 1947–
   Engineering money : financial fundamentals for engineers / Richard Hill and George Solt.
      p. cm.
   Includes index.
   ISBN 978-0-470-54601-7 (pbk.)
1. Engineering–Accounting. 2. Engineering–Finance. I. Solt, George S. II. Title.
   TA185.H55 2010
   658.15024′62—dc22
                                          2010007985

Printed in the United States of America

10 9 8 7 6 5 4 3 2 1

Contents

Preface

There is a traditional gap in an engineer's education. Most academic courses around the world do little to explain that engineering projects depend as much on financial matters as they do on technology. Without money projects don't get built, and without profits there would be no incentive to build them.

In 1996 George Solt pioneered a course on this for senior-year civil engineering students at University College London, and Richard Hill took over the course in 2002. We now teach it to BEng, MEng, and MSc students in civil, mechanical, chemical, and biochemical engineering. When mature engineers hear about it, they all say "Gosh! I wish they'd taught me that when I was a student, I had to learn the hard way." This book is based on that course and aims to fill this gap in engineering education for both students and young engineers in industry. We must be doing something right because, since the first version of this book, our student numbers have grown so much that even though we run two parallel version of this course every year, we've had to move to a bigger lecture hall!

We both graduated in chemical engineering—George in 1950 from Battersea (now the University of Surrey) and Richard in 1970 from the University of Leeds. George spent 35 years in industry, in technical and R&D roles, and a short spell in a management consultant role before becoming a full-time academic. He was also engaged as technical expert for a number of major disputes before the High Court in London.

Richard worked for 25 years in design and proposals followed by 20 years as an independent consultant. We were both in the specialist field of water treatment plant contracting. The experience we gained and the lessons we learned are applicable to all branches of engineering, especially that of project engineering, where the interaction between money and engineering is more difficult and more important than others.

We two have worked together, first in process plant contracting and then as teachers, for so long that we can speak with one voice. When you read "I" in the text, it might be either one of us speaking, but it really makes no difference.

Our experience showed us that there is a need for teaching this subject to everyone who deals with projects. Of course, the need is greatest for engineers, all of whom will sooner or later find themselves acting as either contractor or client in some project.

There's nothing difficult in the bits that make up this subject. So why does it have to be learned? Because in this field everything is interconnected—

that's why the book is full of cross references: The trick is to understand how the whole system works.

To make things easy, we have simplified definitions and explanations wherever we thought we could without actually misleading anyone. On the other hand, you'll see boxes with interesting or entertaining bits that are only marginally relevant, so they're easily skipped if you wish. We hope that they'll help to explain why it's so important for engineers to have a good understanding of financial matters.

Project engineering is, increasingly, an international business and, like most businesses these days it's a 24/7 workplace. I do my consulting work from a small town near London. On a recent project the site team in Shanghai would send me questions by e-mail at 10 AM (which is 6 PM there) expecting to have an answer the following morning. So after 2 PM UK time I could talk to the plant contractors in the United States (who don't get to work before then), and they might let me have the data I need at about 10 PM (5 PM their time) before leaving for home. I would then work through to respond to Shanghai by 1 AM UK time (that's when they start work in the morning): 24/7 is OK provided that you don't have to work all those hours yourself.

So we must be aware of different time zones, different customs, and different currencies. In this new edition we've used U.S. dollars, pounds sterling, and euros in our examples—it's as well to get to used to it.[*]

GEORGE SOLT
RICHARD HILL

---

[*]Exchange rates have varied a lot in recent years, but at the time of writing, the £ and € are roughly similar in value and worth about $1½ each.

Chapter 1

What's It All About?

IT'S ALL ABOUT MONEY

Engineers create wealth. It really is a simple as that. There's an old American tag that an engineer is a man who can do for half a dollar what any fool can do for a dollar. We create wealth by finding cost-effective solutions to problems. Railways, airplanes, atomic bombs, agricultural machines, generation of electricity, mass production of chemicals and pharmaceuticals, computers, and water supply: Engineers have made their mark in every area of human endeavor, and they have done it by reducing costs.

Engineers work in a variety of activities including design, construction, manufacturing, production, research and development, and maintenance—each of which is, ultimately, concerned with money.

ENGINEERING ACTIVITIES

Design is about devising some way of meeting an objective while making the best use of resources—*labor, materials, and energy*—all of which are measured in money. We also have to think of the environment and sustainability. These, too, have associated costs, but we are still learning how to measure them. That leads to difficulties that, to be honest, we haven't yet learned to resolve.

Construction and production are actually the ends of a wide spectrum of making things—a new airport at one end and churning out family saloon cars at the other. At the extremes, construction is a one-off, long-term endeavor that involves at least some novelty, whereas production is continuous and involves little novelty. In between, the two merge continuously into one another. Building a cruise ship is a construction project: Making motor boats is production. The difference is in the size, the novelty, the time scale, and the relationship between the buyer and the manufacturer of the product.

The importance of money matters also changes continuously along the spectrum. It is most difficult and important at the "construction" end. Money

Engineering Money: Financial Fundamentals for Engineers, by Richard Hill and George Solt
Copyright © 2010 John Wiley & Sons, Inc.

is, of course, important in production, but the engineer's work is no more affected by it than if he were making baked beans, and the same goes for those working in maintenance. (So far I've said nothing about research and development, which is quite different, and there is a separate chapter on that subject.)

Construction work is classically undertaken by consultants and contractors. They are the people who are in the front line in the subject of this book—that is why much of it is addressed directly to them. However, most engineers will sooner or later be involved in building, upgrading, replacement, major refurbishment, and the like—all activities that involve consultants and contractors. Disputes can arise because clients don't understand the money problems that contractors face, so there's no excuse for production engineers to remain ignorant.

ECONOMIC ENGINEERING

The history of automotive engineering is littered with technological innovation. Henry Ford's Model T, Vincenzo Lancia's Lambda, Ferdinand Porsche's Volkswagen Beetle, Pierre Boulanger's Citroen 2CV, Alec Issigonis's Mini, and many more. These innovations were driven by economics: to make an automobile that was more affordable but without sacrificing quality and design.

Which of theses innovative automobiles is the most important is a matter of opinion, but they were all far superior, in engineering terms, to contemporary Rolls Royces, which carry the same number of people in large and very expensive gas guzzlers. It's not hard to design and build anything if you can use the most expensive materials, take as much space as you like, and pay no attention to its running and maintenance costs—in short spend unlimited amounts of money.

While these technological innovations were made for commercial reasons, they also resulted in sociological change by bringing automobiles, which had been the preserve of the wealthy, within the reach of almost everyone. Indeed, the Mini became something of an icon of the 1960s, being driven by princesses, film stars, and factory workers. Engineers

> My colleague the keen sales director (see Chapter 7) had a Rolls Royce Phantom III (1937 model) of which he was immensely proud. It weighed 2.25 tons and got 7 miles to the gallon. People often asked whether they might look under the hood. There they would find a downsized version of the Merlin engine—the one that powered the Spitfire and other famous World War II aircraft: a 7-liter V12 engine, with 24 spark plugs and twin magneto ignition (see Chapter 24). "What a beautiful piece of engineering!" they would exclaim. But, of course, it was the exact reverse—it was a classical case of bad engineering. Ettore Bugatti (1882–1947), who knew a thing or two about motor car design, said of the Rolls Royce that it "represents the triumph of mechanics over engineers."

often underestimate the way they affect society both for good and bad (see Chapter 28).

WHO BENEFITS?

Ultimately it's society or "the public." Unless there is a benefit to the general public, engineering innovations fail. Naturally, individuals and corporations make money along the way: Engineers are no more altruistic than other humans. But that is what provides the impetus for innovation. When Isambard Brunel built the pioneering Great Western Railway he did it for the benefit of the burghers of Bristol, who wanted to compete with the London docks for trans-Atlantic trade, and they paid him well for his efforts. But the benefit of lower cost travel between London and the west is still with us.

There are many textbooks on "management." Most, in my experience, are rubbish. I do, however, recommend *Up the Organization* by Robert Townsend. It was published decades ago, but it is still a classic. His answer to the question in the chapter title is "If you can't do it excellently, don't do it at all. Because if it's not excellent, it won't be profitable or fun, and if you're not in business for fun or profit, what the hell are you doing here?"

Businesses exist to make money—that is, a profit—for someone. So, ultimately, do engineers.

WHERE'S THE TECHNOLOGY?

We create wealth by innovations in technology or its application, but good engineers are not, primarily, technologists. We create wealth by using our ingenuity to solve problems, and that usually means using technology. What engineers do is to select and adapt the best technology to get the best fit to the problem, but the most difficult problems are often those of implementing the solution.

The word *engineer* comes from the same Latin word as *ingenious*, which implies that we have to use creative skills and inventiveness to solve problems.

This is where social, political, and above all economic questions get mixed up with pure technology and very often become the controlling element. In my own field of water treatment, we currently have all the necessary technology to convert domestic wastewater (sewage) into drinking water, but persuading the public that the product is safe to drink—a sociopolitical problem—can only be achieved by education and persuasion.

Technological problems usually have a simple right answer—how thick does a 200 mm wide beam have to be to support a uniformly distributed load of 30 tons over a span of 10 m? Engineering problems, on the other hand, have answers that vary depending on the conditions of time and place.

This is illustrated by the history of power generation. You might think that designing the most efficient power station is a fairly straightforward technological problem, but the developments of the last half-century show that this was not the case, and that economics, sociology, and politics all have an influence. By the 1940s, steam turbine generating sets were the main power generators in the world—coal fired in Europe and oil fired in the United States. Their technology had been refined for several decades. At the end of World War II, the nuclear research effort that had produced the atomic bomb was channeled into power generation, which promised unlimited cheap electricity. By the 1960s nuclear power stations were being built around the world.

In the 1980s, while large reservoirs of natural gas were being exploited, it became apparent that the high initial capital costs and massive decommissioning costs of nuclear power made it more expensive than the newly developed combined-cycle gas turbine technology. Moreover, high-profile nuclear accidents such as Three Mile Island and Chernobyl raised such antinuclear sentiment that governments around the world largely ceased construction of nuclear stations.

By the end of the twentieth century, wars in the Middle East raised the cost of oil, which in turn raised the cost of natural gas. In Europe, Russia's manipulation of gas supplies added to concerns about the long-term economic security of fossil fuels. Nuclear power once more began to look cost

Another example from wastewater treatment is wet-air oxidation. Capable of destroying a wide range of organic contaminants in wastewater, the process was developed in the 1950s by F. J. Zimmerman, a British engineer working in Wisconsin. In spite of continuing technological development, wet-air oxidation was always too expensive, in both capital and operating costs, to be attractive for wastewater treatment. In the last decade things have changed: Alternative disposal options such as chemical treatment and landfill, have become relatively more expensive as a result of environmental legislation and taxation. So, half a century after the technology was invented, the economic climate has changed and wet-air oxidation's time has finally arrived.

Geography also has a major influence, as the tale of L & C Steinmüller shows. The company started as a paper mill in Gummersbach, a remote village in the hills near Cologne, Germany, that took its product to the nearest railhead by oxcart. In the nineteenth century it replaced the oxen with an English steam locomotive. It had a multipass fire tube boiler. It failed miserably because, on steep uphill gradients, the boiler water drained back. The tubes at the front rose above the water level and, consequently, burned out. "Ach! We must put the water in the tubes," said Herr S, and had the boiler converted. The water tube design was such a success that boiler-making overtook the papermill as the company's main business. I was told this story when I visited the company works after the war and saw the original machine displayed at the front entrance. "Made in Thetford" I read. "Have any of you gentlemen ever been to Thetford? It's in Norfolk, a completely flat part of the UK: this thing was never designed to go up and down hills!"

effective. Meanwhile environmental concerns about carbon dioxide emissions, together with the threat of carbon taxes, led to a wave of enthusiasm for "renewable" power using wind, tide, and solar energy. Attractive though these technologies are, they cannot meet the ever-increasing demand for electricity. Many environmentalists, including James Lovelock the "father of the Gaia hypothesis," concluded that nuclear power's long-term sustainability outweighed the environmental problems of nuclear waste disposal, which had always been their main concern. Nuclear power seems to be back.

It's not easy to teach this sort of thing. In fact, engineering courses generally just teach technology, which is comparatively easy to teach and easy to assess in exams. But engineering projects live or die by money not technology, and most university courses don't tell you about that. That is why we pioneered this course at University College London and have written this book, which covers most of the outline of the course.

> It's like teaching musical composition. The Juilliard School, the Paris Conservatoire, or the Royal College of Music are very good at teaching composition, but they can't turn their students into Bachs or Mozarts.

WHAT'S A PROJECT?

I've used the word *project*. The best definition I've come across for a project is "something that's never been done before," though we are here concerned only with engineering projects. Whether it's a tunnel under the Channel, a space station in orbit, a bacteria-driven computer chip, an oil refinery, or a bridge from Denmark to Sweden, every new engineering project is different. It needs to be designed and it needs to be built. And it needs to be designed and built economically.

> It seems that when they started to build the famous Sidney Opera House it couldn't actually have been built to the original design. (Civil engineers on the whole are pretty critical of architects for occasionally landing them with this sort of situation.) It was only saved from oblivion by a radical redesign.

The fact that a project is something new means that there must be more or less uncertainty about its outcome. Can we really build it for the budget and in the time proposed? Will it be profitable? Can we afford it? Can we do it at all? There are unknown ground conditions that can affect not only the foundation but the whole approach to construction. Marc Brunel, Isambard's father, had to invent a tunneling shield to construct the Rotherhithe tunnel under London's River Thames. It was innovative, completely untried, and had to be developed on the job. Although it's been updated, the same technology is still used for major tunneling projects such as the UK–France Channel Tunnel.

> The last time I saw one of the shields used to dig the Channel Tunnel, it was sitting on a mound near Dover, with a placard saying "FOR SALE—ONE CAREFUL OWNER."

HOW DO WE BUILD IT?

Most people who want something built don't themselves have the necessary skills or resources to build it. Building something such as an oil refinery requires a multidisciplinary team of engineers—chemical, mechanical, structural, electrical, and civil—to create the design. Then a vast team of skilled builders and fabricators are needed—scaffolders, pipe fitters, riggers, electricians, and so on. An organization is also needed to coordinate their efforts, no matter how small the project.

The construction industry provides these skills and the organization. It covers every scale of construction from the local builder who constructs house extensions to the international corporations that build power stations, chemical factories, and airports. Construction companies get paid to build projects for their clients. Often either the technology or the organization of such projects, or both, need additional skills that a consultant can provide. The agreement between the client and builder is called a contract and the builder is called a contractor. A consultant works somewhere in between them, and there is quite a variety of ways in which that can be organized.

THE CONTRACTING INDUSTRY

It's not just engineers who work in the contracting industry who need to understand its needs. Sooner or later most other engineers (e.g., working in production) will also have dealings with the contracting industry, so they too need to understand how contracting operates as a business. It's much like any other business in its structure and management, but it has many unique characteristics, particularly in the area of finance.

As we'll see later, the contracting industry is very competitive. Only a few large contracts in any particular sector are placed every year, and it is important for a contracting company to win enough of them to survive. This means that profit margins on turnover are low—typically 1.5–5%—although the return on capital invested is quite good (see Chapter 5).

The successful execution of an engineering contract depends greatly on technology and finance; but it also depends on the relationship between the project manager and the contract manager. The first

I worked for a specialist process plant contracting company, and we had sold a water purification system to a large pharmaceutical company. The contract was successful, although there were several disputes during its execution. The following year I was approached by the pharmaceutical company's project manager to see if we would bid for another water purification system at another factory because they were very pleased with the plant we had built. However, he told me that if we were to be given the opportunity to bid we'd have to nominate another contract manager!

is the purchaser's representative, who has to get the project completed and pays the contractor, and the second is the contractor's representative and manages the contract. We will see later how important this relationship is, but first we need to understand a bit more about money.

SUMMARY

- **Engineering is about money.**
- **Project engineering is about risk.**
- **Finance for routine production is similar to that for any other routine business.**
- **The time scale and novelty of project engineering creates different problems.**
- **Every engineer needs to understand about money.**

Chapter 2

Money

WHAT IS IT?

I think of money as the stuff we haven't got enough of—a good oversimplification for a start! Economists probably hate this definition, but it makes an important point: The engineer's job is to meet some demand *by the most efficient use of scarce resources*—labor, power, materials. Our common measure for all these is money—it's a perfectly awful measure but the only one we have.

BRIEF HISTORY OF MONEY

This overview will help to understand where we are now. For thousands of years, people traded by barter—wheat for oil, peacocks for sandalwood, and so on. In time they found that ingots of metal had three properties that made them particularly useful for trading, and so they became the basis for money. Compared with other useful goods, metal was

- **A valuable and scarce resource**
- **Compact and easily transported**
- **Not perishable.**

The Chinese tael was a silver ingot with a standard weight of about 37 g, whose value varied with the price of silver at the time.

At first, the shape of the bits of metal used for trading was unimportant, their value being set by their weight. It became what we now call money when they stamped metal

> In 1941 the Abyssinian currency was still the Maria Theresa dollar. It was originally an Austrian silver coin (28 g) that was minted all over the world until after World War II. Its value was in its weight of silver and mostly was used in northeast Africa. The British army, which invaded the Italian-held country from Kenya, paid their auxilliaries in Maria Theresa dollars and needed great mule trains, their saddle bags weighed down with silver coins, to transport them.

Engineering Money: Financial Fundamentals for Engineers, by Richard Hill and George Solt
Copyright © 2010 John Wiley & Sons, Inc.

into standard sizes, with some mark to indicate a value—but in the beginning that value was still their weight of gold, silver, or copper.

When the value of coins became fixed rather than dependent on their weight, some crafty people found they could snip silver or gold from their edges and melt the snippings down. Milled edges (introduced by Isaac Newton when he was warden of the Royal Mint) prevent that, and our higher denomination coins still have them—even though today's coins are made of cheap alloys.

Carrying precious metal about may be less troublesome than wheat or peacocks, but it is still a cumbersome business and invites robbery. To overcome these two problems, banks (initially in Italy) began to write personal letters promising payment.

A merchant going from Milan to Antwerp to buy English wool would pay cash into the bank in Milan and take with him such a note addressed to the Antwerp branch. He could then travel light and in safety because the bit of paper he was carrying was of no use to anyone else. The Italian bank's agents in Antwerp would pay him in cash when he actually needed to pay for the wool.

In 1695 the Bank of England issued the first *bank notes*, which could be cashed by anyone. These notes were an undertaking by the Bank of England to pay the specified amount in gold or silver. Other banks followed suit. They claimed to have enough in their vaults to cover the total value of all the notes they issued, but quite soon that stopped being true. Everyone really knew that was so, but the notes still represented silver and gold and they were "as good as gold"—just as long as there wasn't a run on the bank. Today's UK £10 bank notes still say "I promise to pay the bearer on demand the sum of Ten Pounds" signed by the chief cashier of the Bank of England. It is a meaningless promise. What can the

If customers lose confidence in their bank, they may all try to withdraw their deposits at the same time. This is known as a "run on the bank." As more customers withdraw their cash, so panic spreads. It can result in the bank collapsing. The financial crisis of 2007 resulted in the first run on a UK bank (Northern Rock) since 1866 when Overend Gurney failed.

From 1450 until 1931 the UK currency included a little £1 gold coin called a "sovereign". It weighs 8 g and is no longer currency, but it is still minted and traded widely for its gold value which (in 2010) is about £150. Assuming the value of gold is constant (it isn't, but it doesn't vary enormously), that represents an inflation rate of 6.5% per annum over the last 80 years.

All other coins were of copper or silver, but then silver coins were withdrawn and replaced by cupro-nickel so that by 1950 there were only a few still in circulation. In 1971, the traditional system of pounds, shillings and pence (£ s d) was changed to metric currency, and all coins were replaced by new ones. We needed that because the big old penny coins (some of which had been in circulation for a century) were now worth more than a penny as scrap.

poor man actually do if anyone tries to enforce it?

The system collapsed after World War I. First the Austrian and German currencies became worthless, and then around 1930 both Britain and America "came off the gold standard"—that is, they

copper. The Royal Mint was having to mint new ones at a loss.

Inflation has soldiered on since then. If the new little "copper" coins were made of copper, they would now be worth melting down, but they're a cheap alloy, not copper.

stopped pretending that their paper money was convertible into gold. For the first time the link between money and a scarce resource had been broken.

After that there was no obvious limit to how much money could be printed. Some people, however, were surprised to find that the more money was printed, the less it was worth. *Inflation had arrived.*

INFLATION

Engineering depends on accurate measurements for which we have units such as grams and kilowatts. Money, our only way of measuring the value of scarce resources, is not only notoriously inaccurate, it is also highly variable. For example, when a book says a ton of sulfuric acid costs $100 it doesn't mean a thing. You need to know when the book was published and how much inflation has increased since then.

Inflation is good news for borrowers because the real value of their debt goes down, so governments, who are the biggest borrowers of all, love inflation. Naturally enough, they aim at the highest level of inflation, which doesn't annoy the voters—the present (2010) UK government has set a target of 2% per annum. The actual rate has recently been a bit lower. For the first time ever, there have actually been complaints that inflation is too low!

Unless the world finds a more sensible means of measuring the value of scarce resources, we can expect that inflation will always be with us. It is now universal, but greater for some currencies than others. There is always the chance that it may get out of control in some places, as it did in Britain in the 1980s when it rose to 16% per annum: It is hard to budget for the cost of a project if you don't know by how much prices will have risen by the time it is finished.

Even then, 16% inflation is chicken feed compared with Germany and Austria in the 1920s, when a postage stamp could cost millions or even billions of marks or crowns, or Zimbabwe in recent years when the price of a loaf of bread would rise during the day.

Governments make up a "basket" of household commodities and regularly measure how much that costs. The basket represents constant value, so this makes quite a good system for measuring inflation. However, it doesn't provide us with a unit to use in everyday calculations—which is where the Mars bar came in.

Measuring the rate of inflation needs some measure of "constant" value. Gold and silver used to serve in the past, but what are we to use now? Some years ago an ingenious journalist claimed that the Mars bar hadn't changed in decades (he was wrong there, actually!) and could be used as a measure of constant value. His examples in terms of Mars bar economics showed how borrowers profit from inflation—which naturally means that lenders lose from it. I myself think the kilowatt-hour (kWh) is a more likely measure of constant value—it seems more plausible, at least.

HOW MARS BAR ECONOMICS MADE SOME OF US RICH

When inflation was up to 16%, it was still possible to borrow money at much lower rates—an incentive to borrow as much as possible. Here's a highly simplified example.

Suppose in 1975 I bought a house for $160,000, paying $40,000 in cash and the rest by borrowing $120,000 to be repaid after 15 years, at 6% per annum interest—that's $7200 per annum, which would come to a total of $108,000 in 15 years.

Suppose inflation between 1975 and 1990 averaged 10% per annum. In 1975 a kWh cost 3¢. Assuming that it has a constant value, its cost in 1990 would then be about 12¢, and the average cost of a kWh during those 15 years 7.5¢.

In 1975 $160,000 was equivalent to 5333 megawatt-hours (MWh), and we'll assume that this is the true value of the house, and that it will remain constant.

In 1990 I sold the house for its true value, which at 12¢/kWh brought me $640,000, so the whole deal works out as shown in Table 2.1.

So I lived rent free for 15 years and made a cash profit of 1560 MWh (=$188,000 in 1990)—which is almost a third of the real value of the house. Table 2.1 is realistic—it represents pretty well how I got a great windfall without doing a hand's turn for it. The next generation is not going to be so lucky.

Table 2.1 Buying and Selling a House

	Dollars ($)	Megawatt-hours (MWh)
Money going out		
Cash (1975)	40,000	1333
Interest (1975–1990)	108,000	1440
Repayment (1990)	120,000	1000
Total Cost	268,000	3773
Money coming in		
Sold (1990) for	640,000	5333
Profit		
Profit	372,000	1560

On the other hand, when my wife's mother was married in 1904, her parents gave £6000 worth of 4% government stock. The idea was that if her husband died, the annual interest income of £240 from it was enough to keep a lady in modest comfort (including employing a maid). Now it will just cover 2 months' worth of the local tax on my little house.

INTEREST

If I offer to give you either a $25,000 car or a check for $25,000, you will be sensible to take the check. You can buy anything you really want with it, including the car. This flexibility makes money more useful than an item of the same value. Rather than buying a car you could hire one, and that means paying rental. But actually cash is more useful than a car, so it seems reasonable that hiring money means you should pay rent for it—it's called *interest*.

Why has interest had such a bad press ever since money was invented? Because people who are desperate for money are obviously a "bad risk," and banks won't lend to them. Then they have no choice but to borrow from those who charge unreasonably high interest

> The Old Testament forbids lending money at interest. By the time the New Testament was written, moneylenders were clearly flourishing, so it seems no one had taken too much notice of the prohibition. Another 500 years later, and the Koran forbids it again.

(we'll discuss *reasonable* interest in a moment). This is called usury and becomes a kind of extortion. Usury often wrecks people's lives, which is why it is rightly considered wicked. But usury is no different from charging unreasonably high rent to people who are desperate for somewhere to live. None of the Good Books mentions that nor the distinction between charging reasonable interest and usury.

WHAT IS A "REASONABLE" INTEREST RATE

The important question is therefore: What is a *reasonable* rate of interest? As with other goods and services, that depends on supply and demand. Historically, however, it is roughly between 1 and 3% per annum for an absolutely cast-iron safe loan. Absolutely cast-iron safe loans hardly exist, so in all real cases we have to think how other influences affect the reasonable interest rate.

Suppose I lend $120 for 10 years to each of 100 people: What interest should I reasonably charge?

The first problem comes from inflation. Suppose a kWh now costs 12¢, so that my $120 is now worth 1 MWh. After 10 years at 2% inflation, the $120 repayments I expect to receive will only buy 0.8 MWh each, so I must charge 2% interest just to offset that loss.

Then there is the risk that some of my debtors will die, go bankrupt, or run off with my money. If I reckon that each year I may lose 2 out of the 100 debtors, I must stick another 2% on the interest to break even.

I'll have to keep records and then chase up my 100 debtors to get paid the annual interest and to get the loans repaid at the end—the cost of the necessary administrative cost cast for that might come to, say, another 1%.

So far we have:

Inflation	2%
Risk	2%
Administration	1%

On that basis 5% interest would only just break even. If I am content with a modest return of 2%, the actual interest rate comes to 7%. It sounds like a lot, but (on the above assumptions) it's *reasonable*. Credit cards have high administrative costs and quite a lot of defaulters, so their interest rates are normally around 15–20% (which, to be honest, I think is verging on usury).

So the variables that govern a reasonable interest rate are inflation, risk, administration costs, and real return.

Borrowing rates are sometimes quoted as APR (annual percentage rate), which usually means the *effective APR*, that is, the compound interest rate.

A simple interest rate of 24% per annum means that you pay annual interest of $24 on a $100 loan. Credit cards often quote rates per month and you might think that a rate of 2% per month is the same as a *nominal APR* of 24% per annum. In fact at 2% per month the effective APR is 27% per annum, and you'd pay annual interest of $27 on your loan.

THE BANKS

As we shall see shortly, loans are a major factor in sizable engineering ventures, so that making loans available is an essential service to them. The many entitles what which perform that duty go under several that names, but I shall call them all "banks." They exist by borrowing money at one interest rate and lending it at a higher one. The Bank of England is the UK government's bank: its "base rate" ("prime rate" in the US) is (more or less) the interest rate at which other banks in the UK can borrow money. One of the Bank of England's main duties is to control the UK economy, for which the level of the base rate is a major tool, and the bank reviews it every 4 weeks.

Every currency system in the world has a central bank of its own that sets a base rate, such as the "Federal Reserve, the Fed, in the United States, and the Central European Bank in Frankfurt.

Projects depend on finance being available when needed. In developing countries you see a lot of half-finished buildings, and the usual reason is that the money ran out, so work had to stop until cash could be found to go on with the work.

A half-finished and uninhabitable house means that money has been spent on something that is producing no benefit, and that's clearly a bad investment. The *benefit* needn't, of course, be in terms of money—for example, the benefit of a new bridge is to make traveling easier, which doesn't happen until the bridge is finished. If banks make money available when it is needed, the project will start to produce its benefit as quickly as possible. *The banks therefore provide an essential service.*

> Why did the great medieval cathedrals take decades and sometimes centuries to build? Maybe the bishopric might have engaged more stonemasons to speed up the work, but often it couldn't. If the work was paid out of the income that the bishopric could raise every year, it probably couldn't raise the money any faster and that limited the speed at which the work could go on.

Bankers have recently sunk low in public esteem—somewhere near politicians and oil executives. But all these perform an essential service without which modern civilization could not function. We don't have to like them, but we need them.

ISLAMIC BANKING

Sharia prohibits the payment of fees for the renting of money for specific terms, as well as investing in businesses that provide goods or services considered contrary to its principles. It's only in the last few decades that Muslims have developed an Islamic system of banking that does not charge interest and so obeys the letter of the law. Some (but by no means all) Muslim banks now follow it.

Any system of banking—that is, one that trades by borrowing and lending money—has two basic rules. First, banks have to make a profit on their service. If they did not provide bankers with a living, would banks exist at all? Second, banks must not lend money without enough security. The recent worldwide economic disaster shows what happens when they break that rule.

The essence of the Islamic banking system is the sharing of profit and loss and the prohibition of usury (*riba*), so banks *invest* rather than *lend*. For example, instead of loaning the buyer money to purchase an item (goods or property), a bank might buy it from the seller and then sell it to the buyer at a higher price to make a profit, but allow the buyer to pay the bank in installments. The bank asks for strict collateral or retains ownership of the item until the last installment has been paid.

This means that, rather than getting a steady but fixed interest income, the bank should share in the risk–benefit from whatever venture it finances. There are indeed a lot of *Merchant Banks*, Islamic or otherwise, which do just that (see Chapter 6). However, classical banking relies on avoiding risk where possible, and we have just discovered to our cost how important that is. The

Islamic system (which is still in the process of being developed) uses various devices that minimize the bank's risk, and so reduce the income they can reasonably expect from their investment. Some of the more common terms used in Islamic banking include *Mudharabah* (profit sharing), *Wadiah* (safe-keeping), *Musharakah* (joint venture), *Murabahah* (cost plus), and *Ijara* (leasing).

A banker friend tells me that in his experience the Islamic system's rules in practice end up doing much the same as conventional methods. This must be true because there have been large projects in the Middle East for which the banks raised money by using both systems in parallel. If the two systems were significantly different in effect, that would hardly be possible.

REAL INTEREST RATES

In fact, banks do not set *reasonable interest rates* as described above but decide in each individual case how much to charge. To put it less politely, they charge the highest interest rate they think they can get away with. In practice, that means it is only competition that keeps the actual rates near the reasonable level. And, because a less-than-totally safe client can be charged higher interest, they are torn between cowardice and avarice.

In each case, therefore, the *risk* factor decides the actual rate. More importantly, it decides whether a bank is prepared to lend money at all.

A well-established company with good security can typically borrow money at base rate plus 1% and its overdraft charge will be base rate plus 2%. Just a whiff of the slightest risk and not only do the interest rates go up but the bank may impose all kinds of conditions. When lending to small companies, for example, they often demand the deeds to the company directors' homes, and so put them at risk of being out in the street if the company fails.

That tells us a bit about what money actually is. Being engineers you now need to know how to measure it.

A banker friend once told me that American banks tend to give in to avarice and the British banks to cowardice. He claimed to have got some system going by which he could make money out of this cultural difference.

To put it another way—the banks refuse to lend to people who really need money, but they'll urge those who don't need it to borrow more. Getting loans depends on appearing safe and solid to the bank. Some desperate games of deception are often played to achieve that.

Also, the banks often impose rules that make it difficult for their clients to move their account. That makes it difficult for the client to take advantage of a lower interest rate a competing bank might offer. It is a barrier to free competition, but there's not a lot that can be done about it. We're getting near usury again.

SUMMARY

- **Money is an inaccurate and variable measure.**
- **Borrowing money means paying interest in some form.**
- **Interest pays for inflation, risk, and administration costs before producing any real return to the lender.**
- **Banks often refuse to lend money if they see a significant risk and will adjust their interest rates to reflect risk.**

Chapter 3

Measuring Money

WHY MEASURE IT?

We measure money because it is among the basic dimensions needed to do anything. All engineering depends on making measurements, preferably accurate ones—which is where money fails miserably.

For a start, inflation makes it a variable measure. If we have to deal with exchange rates between different countries, things are even worse. But even without these obvious failings money is elusive stuff, changing from one form to another, hiding itself only to reappear in unexpected places, and altogether difficult to pin down.

The art of measuring money in all its forms is called accountancy. Only accountants can measure whether a trading venture is (or is not) profitable and sound. However, no matter how good and honest the accountant may be, drawing conclusions from such results always needs care.

> It really is an art. Accountants are often written off as "bean counters" and thought to be dull, boring people. Wrong! They trade in vague concepts and fictions. They are the last poets and dreamers left to us in this modern age.

There are two kinds of accountancy: financial or fiduciary accounting and management accounting.

FINANCIAL ACCOUNTS

Legislation obliges every company to produce an annual set of accounts. These accounts have to be checked or audited by an independent qualified accountant (a chartered accountant in the UK), and are filed by the appropriate government department where they are available for public inspection. They serve two purposes: to determine how much tax the government can extract from the company and to show the public (investors, suppliers, potential business partners, etc.) how the company is doing. The accounts are

Engineering Money: Financial Fundamentals for Engineers, by Richard Hill and George Solt
Copyright © 2010 John Wiley & Sons, Inc.

normally presented in the form of two documents: the statement of financial position (or balance sheet) and the statement of income (or profit and loss account). The actual format of these reports varies from country to country, but the numbers are essentially the same. The statement of financial position shows everything that the company owns (its assets) and everything it owes (its liabilities) at the end of the accounting period (usually the financial year). This actually gives the value of the company at that moment in time. The statement of income shows how much money the company earned by trading during the same accounting period and how much it spent. Taking expenditure from income gives the profit or loss for the period, and this is the *bottom line* of the statement.

Auditors have evolved a set of rules to make the audit as reliable as possible, but it is far from an exact science. Naturally enough, the company's aim is to get the audit to minimize its tax bill, while presenting the best possible image to the world—two purposes that are mutually contradictory. Doing it within the law can be quite a challenge. That is why departures from the law—some great, some small—are not unknown.

When such departures occur, it is hard to work out just exactly what the truth should have been—sometimes the fraud is so complex or ingenious that we never get to the bottom of it. However, as we shall see, accounts can be massaged without breaking the law. It is not always easy to know just where lawful massaging stops and fraudulent accountancy starts.

Auditors have a hard time because they serve two masters: On the one hand the law demands a true and unbiased picture—if auditors do not follow the regulations, they will lose their license. On the other hand, they get paid for their services by the company they are auditing, so they have to please their paymaster or risk losing the business. It's not surprising that auditing has developed a range of practices, from slightly dodgy to completely fraudulent. Several enormous accountancy-based frauds such as the dawnfall of Enron have recently become public, but they're just the tip of the iceberg.

> Anyone who thinks accounts are boring should read *The Smartest Guys in the Room—The Amazing Rise and Scandalous Fall of Enron* by Bethany McLean and Peter Elkind.

Audited accounts are presented with a great show of precision, but as many of the "measurements" that accountants make are really guesswork, the result always carries an element of doubt. There is a lot of freedom in setting up an audit, which means that even accounts that are perfectly legal may be (and often are) massaged to give a more or less distorted picture. Here's an example of how an audit can be fudged without breaking the law.

PROVIDING FOR THE (UN)KNOWN

At the end of a contractor's business year some contracts will be partly finished and others finished but still under guarantee. All these carry some risk of costing money in the future, so provisions must be set aside to represent the foreseeable risk. This is purely a paper exercise based on a guess at the probable loss of profit that can be expected. The auditor cannot make that judgment alone: the company will consider what it thinks is reasonable and the auditor will accept that figure unless he or she deems it unreasonable.

> *Provisions* are a guess at a likely future loss. The most obvious example: When banks publish their audit, the provisions are made up mostly of loans they think may not be repaid, multiplied by the probability of that happening. Or, a company that is in the middle of a lawsuit should make provision against the cost that would result from losing it. Whatever they are for, provisions are always going to be guesswork and therefore liable to massaging.

Provisions count as a loss and reduce the year's profit. So if a company has had a good year, it will want a high estimate for the year's provisions for two reasons. First, these "losses" serve to reduce the profit and so reduce the company's tax bill. Second, if the provisions are overestimated, they give the company a hidden reserve to be carried forward into the next year. Next year, when it turns out that the provisions were overestimated, there is a surplus, which goes into the year's income and increases the profit. By that time, there is another audit and a fresh set of provisions to be considered. By overestimating

> It is important to realize that this is a paper exercise that is done purely for the audit. It is not the same as hiding a pile of banknotes under the bed. No real money is put aside for the expected losses.

the provisions year after year, the company can provide a virtual permanent reserve for a rainy day.

If, on the other hand, the year has been bad, underestimating the provisions increases the profit reported by the audit, and that helps keep shareholders and the bank happy. As long as the individual provisions for each contract are reasonable, it's quite legal. This game is played every year, so a single year's audit must be expected to contain some such hidden sums, but in the long term it must all come out in the wash.

BRANDING

Akzo Nobel's paint brand Dulux has acquired a worldwide reputation for consistent high quality. Specifications for engineering work, therefore, often say that "painting shall be done using Dulux or equivalent quality." Years of consistent advertising (with several generations of Old English Sheepdogs appearing in Dulux advertisements) have helped create this brand image, so the reputation is partly thanks to serious investment — a type of investment just as real as investment in bricks and mortar. If Akzo Nobel were to sell the paint business, the buyers would have to pay a huge sum for the brand name. This kind of asset is called *goodwill*. To show how valuable goodwill can be, after Rolls Royce aeroengines had split from Rolls Royce cars, the brand name Rolls Royce and its distinctive badge (as applied to cars only) were sold to BMW for several million British pounds.

Any company that has built up good relations or a good reputation has acquired some goodwill, but it does not have to appear in the audit unless there is some good reason for it. If it does, the problem is how to value it. A minimum valuation could be the cost of past publicity (Old English Sheepdogs don't come cheap), but that comes nowhere near its potential sales value in either of my two examples. The value that has been given to goodwill in an audit is always a guess. That makes legal fudging possible, but there is also room for fraud.

These are just examples of a variety of items in an audit that are largely guesswork and can therefore get fudged. In the end any fudge will come out in the wash, so it takes at least three consecutive years' worth of audits for a reasonably reliable look at a company's finances.

Everyone should understand what the main items in a set of published accounts mean. Lots of books teach basic accountancy but, sorry, this isn't one of them. This chapter only highlights why it is important to know something about accountancy, but you'll find a short guide to accounts in Appendix 1.

MANAGEMENT ACCOUNTS

Management accountancy, on the other hand, is quite separate from auditing and has a completely different objective. Management accountancy is for telling the company's management what is going on, so cosmetic exercises are not only pointless but can be harmful.

Whereas the audit has (by law) to be done by a licensed outsider, management accountants are company employees. The law says nothing about their qualifications, but in all but very small companies this is an important job that needs well-trained people.

Our chief executive officer (CEO) used to estimate the total annual cost of the directors' bonuses in advance and add one twelfth of the sum to the monthly *salary* figures in the management accounts. I suppose the idea was to conceal from the staff that we were paying ourselves a bonus. As this bonus, like our salaries, was quite modest, I didn't see the point. Anyway, the management accounts people must have known, so it wasn't a secret even within the company.

Management accounts set out to measure the real state of affairs as truthfully as they can, which means it is confidential company stuff. It tells the company management how they're doing in general, whether individual contracts are going wrong, and what action may be needed. Speed is essential, so good information technology systems are absolutely vital. However good the management accounts might be, if, they turn up after the event, they are likely to be too late to be helpful. On the other hand management accounts don't have to be absolutely exact, just as long as they lead to generally correct conclusions.

> One of my former CEOs used to move money in the management accounts from contracts that were doing well to contracts that were losing money. This had the effect of making the better managed contracts look worse than they were, and the worse managed contracts look better. It may have made the managing direction (MD) feel that he was running a sound company, but it didn't fool anybody and, in the end, the company went bust.

Generally, the tools used in management accountancy are similar to those used in financial accountancy and center on a balance sheet and a profit and loss account. However, the object of the exercise is different, and management accounts are usually more detailed and show the actual cash situation.

One more point: fiddling the financial accounts is illegal; fiddling the management accounts is just stupid. It is one of my qualifications for writing this book that, during my 30 years in the contracting industry, two companies failed while I was a director. One of them only went bust because the management accounts of a subsidiary company were fiddled with—not with criminal intent so much as through wishful thinking (see Chapter 4). They fooled the management into thinking everything was going well when it wasn't.

SUMMARY

- **Stop thinking of accounts and accountants as boring.**
- **Learn how to read what they seem to say.**
- **Recognize where they might be misleading.**
- **Management accounts must be prompt, informative, and should not mislead.**
- **The accounts department needs cooperation from everyone in the company to achieve these objectives.**

Chapter **4**

How Things Can Go Wrong—1

This is the story of a very small contracting company, newly formed to work in a branch of environmental technology that was then quite novel. A little over 50% of the shares were held by a somewhat larger parent company, but as its location was quite a long way across the country, contact between the two was not particularly good. The MD of the parent company would visit from time to time, but there wasn't much contact apart from that. The rest of the shares were held by some of the staff of the company, especially its managing director, a well-qualified engineer whom I shall call Jim.

Jim was a charming man, all energy and optimism. In the first 2 or 3 years under his leadership the turnover increased steadily. Its audited profits were extremely small, but that was only to be expected of a new company—it would not even have been particularly surprising if it had made small losses instead.

Then, after some time, the parent company began to notice that Jim's overdraft had grown and grown. The bank allowed it to do that because the parent company guaranteed it. That is why no serious alarm bells had so far rung, but now the figures had grown so large that they gave serious concern. As luck would have it, the parent company's MD had had a heart bypass operation and was off work for a long time. So no one from the parent company had visited for a while, and now no one could be spared to look into the affair. An outside management consultant was engaged to do so.

The outside consultant realized at once that something was very wrong, but it took him some time to discover the full extent of the problem. When he reported back to the parent company he said: "When I opened this can of worms, the worms just kept on and on flowing out. There is no way they could ever be stuffed back into it." The deficit that Jim's company had incurred was far worse than it at first appeared.

The disaster was due to Jim's energy and optimism. He took on more and more contracts in a technology that was not properly established and in which no one really had much serious experience. To get the contracts, he had had to undertake guarantees of performance, and his installations frequently failed to meet the guarantee. Someone (often Jim himself) then had to return to the site to assess the problem and, as often as not, undertake some remedial work.

Jim would return from these outings and tell his accounts clerk that the cost of the visit and of the remedial work was an extra on the contract, and the client was going to be charged for it. The clerk (who was just that—a clerk, not a qualified accountant) did as she was told. Some of these "extras" were actually invoiced. The clients had merely insisted that Jim's company should meet its guarantees, so they did not pay these invoices, in which they were perfectly justified. Those "extras" that were not invoiced were entered in the books as work in progress. All of these costs therefore appeared in the audit as trade creditors, or as work in progress, and were classed as assets. That was how the company's apparent profit was created, and why it had an ever more threatening overdraft.

With good contact between the parent and subsidiary companies, the problem would have been recognized earlier. But with the MD of the parent company out of action at the critical time, the situation became irretrievable. He recovered, but the company didn't. It very nearly pulled the parent company down with it: It had to be rescued by a takeover.

SUMMARY

- **A company that operates without an adequate accountancy system can run into desperate trouble without anyone noticing.**

- **In those circumstances, too much energy and optimism, plus a measure of self-deception, leads to ruin.**

Chapter 5

Good Company

LIMITED LIABILITY

Almost all trade is carried on by organizations that are companies, so it's as well to know what a company is and what it can do.

When people get together to form a trading organization, they can do so as partners. The drawback to that is that if the partnership fails, each of the partners is liable for all the debt that it has incurred, however great that might be. The idea of a *limited liability company* originated in the seventeenth century in order to avoid this kind of personal disaster.

A limited liability company has a persona—that is, in law it is treated as a person, with rights and responsibilities of its own. On the other hand, its backers (now called *shareholders*) are protected from everything except the loss of their original investment.

The novel idea that you could join a trading venture that did not carry the risk of unlimited losses if it were to fail was obviously attractive. It seemed to offer a safe way to riches, and a kind of mass hysteria gripped London in 1720. Companies were floated for the most absurd purposes, including the famous one for a project "yet to be revealed." Shareholders found that even their limited losses could bankrupt them. We had another example of mass hysteria in 2001 when investors rushed to buy shares in dot.com companies—that is, those proposing to use the potential of the Internet. Once again, the losses were heavy.

Every company is run according to its own rule book, which is called the *articles of association*. These rules can only be changed by a vote at a general meeting. When a company is first formed, it must have some nominal capital, though it can be a very small sum. It may also have shares that have not yet been issued. At any time after that, the shareholders can agree to create and issue more shares and sell them in order to get more capital into the company.

In a company, the shareholders own fractions of the company that are proportionate to their respective shareholdings, with voting power in proportion. The *nominal capital* of a limited company is the sum of the nominal

Engineering Money: Financial Fundamentals for Engineers, by Richard Hill and George Solt
Copyright © 2010 John Wiley & Sons, Inc.

face value of its shares. When first issued, the shares must not be sold for less than this nominal value, but they may be sold *at a premium* for more. After that, the nominal value of the shares is of little real importance; they can be sold for more, or less than that, or become worthless.

Normally, the shareholder can only turn an investment back into cash by selling the shares. If the company has done well, the price should be more than the shareholder paid for it, and vice versa. If the company goes bust, the shares are worthless, but that is the maximum that a shareholder can lose—hence the limited liability bit.

People invest in shares in the hope of getting a good dividend (a distribution of part of the company's profit) and/or owning shares that can be sold at a profit. On the other hand, the company may not do well, and in that case shareholders will lose out. Shares always carry some risk.

The UK's electricity transmission network was privatized as National Grid plc. It went through some merger deals that caused the accountants to change the nominal value of the shares, as a result of which it is now $11^{17}/_{45}$ p (eleven and seventeen forty-fifths pence—I'm not kidding). If the nominal value were important such an absurd value wouldn't be acceptable.

A grim example of the risk in shares: I was trustee of a charitable trust that had quite a lot of cash. Trustees have a duty to invest safely, so we put half of it into shares in Baring Bros—Her Majesty the Queen's personal bank. You could hardly think of a safer investment, but it had a trader called Nick Leeson who misused the bank's lax controls to gamble disastrously on the futures market and effectively bankrupted the company. Its shares became worthless, and the whole company was sold to a Dutch bank for £1. The Dutch repaid all the loans but the shareholders got nothing.

PRIVATE AND PUBLIC COMPANIES

When four colleagues and I set up a contracting company, it was as a *private limited company*. This describes a company whose shares are not freely available to the public investor and, in the UK, have the letters *Ltd* after its name. Their articles of association contain some set of rules to ensure that the shareholders control who may buy the shares. The commonest of these says that any shareholder who wants to sell shares must find a buyer, agree on the price, and then offer the shares to the existing shareholders at the same price. This is the sort of arrangement that ensures that the shares of a family company can be kept in the family.

It is very hard to value the shares in a private limited company. The minimum price is what the company would fetch if it were broken up and its assets sold. However, if the company is trading successfully, that is obviously much too low. There isn't a maximum price. If the prospective buyer wants the shares badly enough, he or she might pay some absurdly high sum.

Companies that are big and well established enough to get accepted by a stock exchange (and there are fixed rules about that) are called *public limited companies* (designated *plc* in the UK), and their shares can then be sold freely on the open market—that's what a *stock exchange* is for. Britain alone used to have dozens of stock exchanges, but now there are only two sizable ones left: The London Stock Exchange and the AIM (Alternative Investment Market) exchange. And, of course, there are dozens of other stock exchanges around the world.

When Ltd companies have done well and gotten big enough, they often *go public* by converting the company to a plc. One object of that is to make it easy for the existing shareholders to sell their shares and so cash in on the progress the company has made. The shares effectively become worth more if the company becomes a plc because they are easier to turn into cash. Another objective of going public is that the stock exchange can help companies to raise more capital.

On the other hand, the company loses control over who owns its shares. They could, for example, be bought by a competitor or, as often happens, be taken over completely by a *hostile takeover bid*. Size isn't everything. The international confectionery firm, Mars, is still a private limited company owned by the Mars family—you couldn't take them over however hard you tried. But, while we're talking chocolate, Cadbury (a plc originally started by the Cadbury family) is currently the object of a takeover bid from Kraft with Hershey and Nestlé waiting in the wings.

> I have shares in a small but successful Ltd company with unusual specialized know-how. A huge company in another country (which doesn't have this know-how) is trying to take over the whole business as I am writing this. We would rather stay independent. We wish we knew why the prospective buyers are making their approach. If it is as a straightforward business transaction, we will probably not agree on a price. We suspect, however, that their government is behind this, which means they may be getting some kind of subsidy, and they might offer a very large sum. Every man has his price. …

> Another grim example of risk, this time in government bonds. My family comes from Vienna. My grandfather died at 80, just after the end of World War I. When my father opened the family company's safe, he found that the old man, a good patriot of the Austro-Hungarian Empire, had invested every cent he had in government war loans. Because the Austro-Hungarian Empire had lost the war and been dismembered, these were now worthless. My father's memoirs record the sour pleasure he got from dropping the loan certificates, one sheet at a time, from the fourth floor window and watching the family fortune flutter into the street below. Moral: Nothing is totally safe (not even a stockpile of Mars bars—the rats might get at it). The important thing is to weigh the risk of any investment, and balance it against the return it offers.

There is no logic in buying shares unless they promise a better return than safer investments such as government bonds and savings accounts, which currently (2010) yield about 4–5%.

The share price of a public company reflects how the public sees the company's future. Shares may give a poor return purely in terms of dividend if much of the profit is retained to invest in company growth. Investors buy their shares if they think there is a prospect of improved future trading, with larger dividends and the possibility of selling the shares at a profit. Shares in private companies have no share price in that sense; each deal is negotiated.

WHO RUNS THINGS?

A company is owned by its shareholders, but they have no duties in terms of day-to-day management. This separation of ownership and management means that, particularly in large organizations, management's self-interest may not necessarily coincide with that of the shareholders. This has focused attention on corporate governance, which is intended to ensure that the interests of all the stakeholders in a company—the shareholders, directors, management, and employees—are fully represented. Many larger corporations have a two-tier corporate hierarchy. The first tier is the board of directors or governors and the second is the senior management, that is, individuals hired by the board. Most of what follows refers to public limited companies.

BOARD OF DIRECTORS

The board of directors or governors is theoretically elected by the shareholders to run the business on their behalf, but, in practice, it tends to be a self-perpetuating oligarchy. The role of the board is to monitor the managers of a corporation, acting as an advocate for stockholders: The board of directors should ensure that shareholders' interests are well served.

There are two types of directors: internal and external. The first are individuals chosen from within the company—typically the chief executive officer (CEO), chief finance officer (CFO), and chief operations officer (COO) will be included, but a director can be a manager or any other person who works for the company on a daily basis, often a representative of the shop floor workers or trade unions.

External directors have the same responsibilities and voting rights as the internal directors in determining strategy and corporate policy, but they are not part of the executive management team and are therefore considered to be independent from the company. They are there to provide unbiased and impartial perspectives on issues brought to the board.

A *merchant bank* invests its own capital in a client company. An *investment bank* raises capital for its client company from external sources.

Often they represent a major corporate financier such as a merchant bank or investment bank or because of some special influence—politicians frequently take up such positions.

The titles given to various positions in companies vary between countries and companies, but the structure is usually similar. The board must present its annual audited accounts to the shareholders, usually with a company report, and convene an *annual general meeting* (AGM) of shareholders. If some major decision has to be made, the board or a sufficient percentage of shareholders can also convene an *extraordinary general meeting* (EGM). At these meetings the shareholders have the right (but no obligation) to vote on any resolutions put to them. They also vote to appoint or sack individual directors, and if they're really unhappy, they can sack the whole board.

The directors may be, but don't have to be, shareholders. The board of directors is controlled by the chairman of the board who is elected from and by the board and is responsible for running the board smoothly and effectively, although decisions are supposed to be made by a majority vote of the directors. His duties typically include maintaining strong communication with the CEO and high-level executives, formulating the corporation's business strategy, representing management and the board to the general public and shareholders, and maintaining corporate integrity. Frequently, particularly in small companies, the chairman is also the CEO, but this practice is not considered good corporate governance because of possible conflicts of interest.

The directors are officers of the company, as is the company secretary who may or may not be a member of the board—often the finance director, or CFO, does this job. This person is responsible for making sure that the company stays within the law and maintains proper record books and accounts.

SENIOR MANAGEMENT TEAM

The day-to-day running of the company is under the control of the CEO who reports directly to the board of directors and is responsible for executing board decisions and initiatives and for maintaining the smooth operation of the firm with the assistance of senior management. The CEO will normally be a board member and may be designated as the company's president or managing director (if not the chairman). The CEO wields tremendous power in a company's corporate structure. Although the board of directors has the ultimate say in most matters, its power is focused into the CEO. This concentration of power is good when the CEO uses it to improve the company, but a bad CEO can use this same authority to run a company into the ground.

The COO is responsible for the corporation's operations including marketing, sales, production, and personnel. More hands-on than the CEO, the COO looks after day-to-day activities while providing feedback to the CEO. The COO is often referred to as the senior vice president.

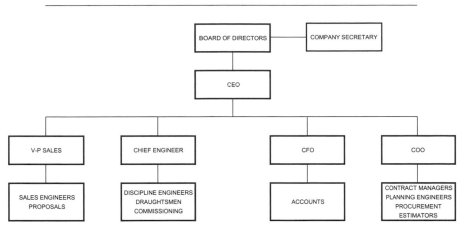

Figure 5.1   Typical contracting company organization chart.

The CFO also reports directly to the CEO and is responsible for analyzing and reviewing financial data, reporting financial performance, preparing budgets, and monitoring expenditures and costs. The CFO is required to present this information to the board of directors at regular intervals and provide this information to shareholders and regulatory bodies such as the U.S. Securities and Exchange Commission (SEC). Also usually referred to as a senior vice president, the CFO routinely checks the corporation's financial health and integrity.

In a medium-sized contracting company, there will be a senior manager or vice president for each particular area of activity, and the company structure may look something like the organization diagram in Figure 5.1.

SHARE PRICE

Once a company has been formed, the nominal value of the shares becomes unimportant, although it still appears on the balance sheet. For buying and selling shares in a plc, the share price, like that of anything else, is a function of supply and demand.

A good way of assessing how a plc is seen to perform is the ratio of the share price to the company's earnings per share: this price/earnings ratio is listed in the financial

The Northwest Company Fund is a centuries-old Canadian company that owns many local shops from Alaska to Greenland—very reputable, very steady. One day in 2009 its share price dropped sharply for no visible reason, until you discover that several million dollars' worth of its shares were traded that day. It looks as if some big shareholder had put them on the market, so the price dropped. Two days later the price was back to its previous value. The company's trading position hadn't changed at all.

columns of the press and on the Internet as the *P/E ratio*. The shares in run-of-the-mill companies show a P/E of around 12, which is equivalent to a return of 8.5%—quite a lot more than one could get on a "safe" investment such as government loans, but as I said, investing in companies always involves some risk.

This percentage is not what the shareholder receives annually in terms of dividend: that's usually much smaller because the company will usually plow back a part of its

If the P/E is much lower than 12—say 9 or less, this means the public thinks the company is not expected to do well in future, so the share price has fallen.

By contrast shares in Smith & Nephew plc have a P/E of over 20. That's because it has excellent technology in artificial knee and hip joints. As people in developed nations live ever longer, the demand for these replacements must rise. The expectation therefore is that the share price will go up in the future, so it's worth paying over the odds now.

earnings to increase its capital. When we founded our company, we made a policy decision to divide each year's profit (after tax) into three—one to pay staff bonuses, one to the shareholders as dividend, and one to be plowed back and stay in the company. It seemed both fair and logical.

Plowing profits back into the company adds to its wealth, and so the value of the shares will eventually rise and the shareholders will get the benefit.

Very occasionally the shares of a plc get into the hands of one or a few owners with a sufficiently large majority to entitle them to decide to convert it back into a private limited company. Richard Branson recently did that with Virgin Airlines.

The two types of share ownership make a great difference to the

Yale's Robert Shiller has a website that shows how the average P/E of us shares chuntered along at 10 ± 5 for most of the twentieth century, but over the final few years rose steadily, and by 2000 it averaged 20. This meant in effect that investment in shares yielded a return of 5%, which—given the risk of failure—is too low. Sure enough, share prices crashed 2 years later and then settled at a lower and more sustainable level. In 2008 they crashed again as a result of the banking crisis.

character of a company. The board of a private limited company (or their friends and family) usually own most of the shares, so they are working for themselves. The share price is normally unknown and unimportant because no one is buying or selling them. If the company does well, the owners can pay themselves bigger salaries and bigger dividends out of the profits. All such matters are discussed and settled in the privacy of the company itself.

In short, in a private limited company, if the company performs well, the directors benefit directly. Thus, they concentrate on improving its performance, and that makes a difference to the atmosphere of the whole company.

By comparison, the quoted share price and its daily movements are crucial to a plc. It is not only a measure of the public's confidence, but if investors find the share price falling, they may get troublesome at general meetings. Equally important is that the directors are often paid with issues of free shares, so they have a personal interest in jacking up the share price. In a plc, therefore, self-interest motivates the board to raise the share price by whatever means come to hand. That also makes a difference to the whole company's atmosphere.

That's why the company report that accompanies the annual audit of a plc always contains a fair amount of window dressing. They generally promise a better future, even if that is doubtful. Any statistics of performance over the past years will be carefully selected to start from the most favorable base—the one that shows present performance in the best possible light.

In the past shares were generally bought by the wealthy as a long-term investment for their old age. When they needed cash they would sell the shares expecting to see a large increase in value as the company prospered by plowing back profits. Margaret Thatcher's monetarist policies of the 1990s resulted in many more small investors buying plc shares in the hope of getting a quick return in the form of dividends. This meant that paying dividends rather than plowing back profits increased the attractiveness of a share and, hence, its price. As a result many companies in the UK failed to invest sufficiently for the future and, when times got hard, they failed.

It's an interesting fact that the average tenure of CEOs in big UK plcs is about 3 years. This means that a CEO will benefit from making everything look wonderful at the moment when he or she leaves the company: They can sell shares at a high price, get a nice bonus on leaving, and leave a successor to discover that things aren't nearly as rosy as they looked. The successor, in turn … etc. etc.

SHAREHOLDERS

Together, management and the board of directors have the ultimate goal of maximizing the shareholder's investment. In theory, management looks after the day-to-day operations, and the board ensures that shareholders are adequately represented, but the reality is that many boards are made up of management.

The shareholders should ensure that there is a good balance between internal and external board members and separation of CEO and chairman roles. There should also be a variety of professional expertise on the board from accountants, lawyers, and executives. It's not uncommon to see boards that are comprised of the current CEO (who is also chairman), the CFO and the COO, along with the retired CEO, family members, and the like. This does not necessarily signal that a company is badly run, but a shareholder might question whether such a corporate structure is in his or her best interests.

The directors of public corporations would rather have a large number of small shareholders than have the voting power concentrated in a few hands—it makes shareholders' rebellions much less likely.

When such companies want to raise new capital, they can issue new shares that are offered at an attractive price, to make sure the public buys up all the shares. As a result these issues are normally oversubscribed—there are more offers for shares than there are shares available. Sometimes they are only offered to existing shareholders (a *rights issue*) but the same is still true.

There are exceptions to everything, and here is a recent one of a company that reckoned it had too many shareholders. O2 plc was originally created when the nationalized telephone industry was privatized, and all its employees had the chance to buy shares at a very good price. The company took the opportunity of a major reorganization of its shareholding to offer to buy out a huge number of shareholders with tiny holdings. These might be worth as little as £20: Sending all of them the accounts and company reports, and paying them minute dividends twice a year, was uneconomical, the company reckoned.

The company then creates rules for issuing the shares, by which applicants for small parcels of shares get the number they requested in full, while larger requests are progressively scaled down. Result: a large and docile flock of sheep as shareholders and an easy life for the board.

SUMMARY

- **The limited liability company is the most convenient form for most trading ventures.**
- **It starts by setting up its articles of association and selling shares.**
- **If it's big enough it can apply to a stock exchange to become a public limited company (plc), which makes its shares available to the public.**
- **Investors buy shares in the expectation of getting a better return than from some safer kind of investment but take the risk that the company might fail.**
- **The share price is important to a plc, but a private company doesn't really have one.**

Chapter 6

Capital

WHAT IS IT?

Capital is an *accumulated* sum of money that is invested for *long-term benefit*. Money spent on anything else is *current expenditure*.

By *accumulated* I mean that it is somebody's savings. The money may be your own (or a company's own) savings or your friends' and relations'. Or it is borrowed from banks, which got it by borrowing other people's or companies' savings. Governments can, of course, appear to create money out of thin air, but to explain that would bring us into the realm of economics (Thomas Carlyle rightly called it "the dismal science"), which I avoid like the plague.

Buying a house is a capital outlay made for the long-term gain of living rent free. You typically borrow the money from a building society, in the form of a *mortgage*—a loan based on the security of the house. The money comes from people who have put their savings into the bank's savings account. Money spent on a holiday is not a capital expenditure—the sun-tan fades quite quickly.

Long-term benefit is a bit vague. It depends, among others, on the scale of the expenditure. A good copier-printer is a capital outlay to a one-man company but a current expenditure for a large one that will have to buy dozens of them. So

Nature is full of examples of long-term investment—an oak tree labors all summer to produce acorns, which are full of nourishment. That nourishment is the product of surplus energy, which the tree has produced and is now devoted to helping the seedlings to survive, take root, and perpetuate the species.

Whether an expense is classed as capital or current expenditure is important in accountancy. Current expenditure comes out of the company's income, reduces the profits, and therefore the tax bill. That's good news, except after a disappointing year's trading, when the company would like to show a healthier profit in its accounts. Then the border is shifted to minimize current expenditure and increase the paper profit. Who says accountancy is either boring or exact?

Engineering Money: Financial Fundamentals for Engineers, by Richard Hill and George Solt
Copyright © 2010 John Wiley & Sons, Inc.

there is a gray area when the accountants can decide under which heading the cost should be shown.

These are all my own definitions. They are oversimplified to make things easier to understand, but they do serve our immediate purpose.

All ventures need capital of two types: *fixed capital* and *working capital*. The easiest way to understand this is to look at a simple example.

WHAT'S IT FOR?

Suppose you want to start a corner shop. First, you have to pay for the premises (or a lease), the furnishings, shop window, chest freezer, and shop sign. These things are not readily turned back into cash again, so they're called *fixed capital*.

Then you need money in the till, and more in the bank for paying wages, bills, and the like. You have to buy the stock—newspapers, magazines, candy, and so forth—all of which you hope to sell. That's *working capital* because you hope to turn it into cash soon.

> Think of capital as being like a public water supply undertaking—the water works, pipes, and pumping stations are fixed capital, and the water in the reservoirs is the working capital. Without the necessary amount of water in the system, the pumps run dry.

This distinction is common to all trades. In manufacturing most companies have fixed capital invested in premises and machinery. Contracting, however, is quite different.

The most primitive type of contracting, for example, is the itinerant builder who knocks on the door and offers to repair your roof. He too needs to invest capital before he can get any work: He has to acquire a few tools and his battered old pick-up, and has to feed himself until he gets paid for his work.

When he does get a job, the first thing he will do is to demand cash in advance with which to buy materials—he hasn't enough capital in his business to do that, so he has to use his client's. This shows a typical characteristic of all contracting—it requires relatively little fixed capital. Offices, machinery, cars, and the like can all be rented. Much of the workforce is hired for specific contracts—not only the people working on the site, but, for example, freelance or agency staff taken on for each specific job.

On the other hand, contracting does need a lot of working capital: The contractor has to pay his employees, buy materials, hire machinery, and so on, while the contract proceeds. It is normal for contractors to be unable to cover all these costs unless the client pays something before and/or during the contract period (see Chapter 9).

While working capital is a crucial factor in all kinds of trading, it is particularly important in contracting. Actual needs vary widely not only between different types of technology but also within them. For my examples I will assume that contractors need $1 working capital for every $5 of annual turnover—this 5:1 ratio is a useful average. All other things being equal, the ratio of turnover to working capital is a measure of how effectively a company is managing its business.

A company needs working capital with which to pay its bills as they come due. If it trades at a rate greater than its working capital can support, it will, sooner or later, find itself in a squeeze when it can't pay its bills. It might be running profitably, and it might be expecting some large payments quite soon, but neither is much help (see Chapter 8).

Companies must have capital of their own to start with, and if they trade profitably, they will plow back some of their profit into the company to increase that. In addition, most of them rely on bank loans to supplement their working capital so that they can achieve a turnover higher than they could sustain using only their own cash.

Working capital can be boosted with bank loans, but banks look for security and for fixed assets to guarantee it, before lending money. They reckon logically enough that when a company fails, it is because there isn't any cash left, but the fixed assets will still be there. The terms on which they have granted the loan usually give the banks first rights over those, and they can sell them off to get back at least some of the money they lent.

While typical manufacturing companies have fixed capital in the form of factories, machinery, warehouses, and the like, most contractors don't have much in the way of fixed assets to serve as security. Thus, the banks won't lend them a lot of money. As a result, for many contracting companies shortage of working capital is what sets limits to their turnover.

It's not quite true that banks have first rights to a company's assets. By law the government has first rights over the remaining assets of a failed company to recover any tax the company owes. In fact, it is very often the inability to pay their tax bill that pushes companies over the edge. Another thing: Trading when knowingly insolvent is a criminal offense and, if a board of directors does that, members are personally liable and could go to jail. I can report that when I was in that position for a period of some weeks, I found it a most disagreeable sensation.

WHERE DOES IT COME FROM?

The common ways to get capital are by selling shares in your business or by borrowing it from a bank.

Selling shares means that you actually sell part of the company. People or organizations buy shares because they offer a good prospect of return on their money from dividend payments or from a rise in the value of the shares, or both. The offer has to give them that prospect. A bank also looks for a good return when lending money but will settle for a smaller return if it is reasonably satisfied that it is likely to get its money back.

Selling shares means there will be more shares on which you have to pay dividend, if you pay any at all. A loan means you have to pay interest to the bank regardless of whether you have done well or not. Either way it will cost the company something. The difference is that dividends are supposed to be paid out of profits the company has made, but interest must be paid in both good and bad times.

> There's an intermediate kind of share issue called *preference shares*. They carry a fixed dividend whose payment comes before that of an ordinary dividend, but in bad times the company can skip the preference shares' dividend as well. They are rather out of favor these days.

RAISING CAPITAL BY SELLING SHARES

The best way to explain this concept is to describe how, many years ago, we set up our own company. The figures are invented, but this is what happened.

An ex-colleague called Chris recruited me and three others to found a process plant contracting company in our highly specialized field. We each declared how much we could personally invest (I myself took out a loan from my bank to raise the cash). To avoid the fuss of starting a company from nothing, we bought a dummy limited company from a firm of lawyers. Its nominal capital was £100, and it had never traded. It was called Middle East Trading Co Ltd, probably for some historical reason. We had no thought of trading in the Middle East, but that didn't matter. Law companies set these companies up when they've nothing better to do and keep them in stock for just this purpose.

Chris then sold the idea to a merchant bank—that's not a bank where you have your current account, but one whose business consists of making investments of this kind. He managed to make the bank sufficiently confident of our prospects to put up quite a lot of cash in order to take up shares, so that altogether we had £140,000 to start the new company.

Having bought the £100 worth of shares, we were the shareholders and entitled to call an official extraordinary general meeting. The five of us plus a man from the merchant bank appointed ourselves as the board of directors, changed the name of the company, increased the nominal capital to £140,000, and between us bought the newly issued shares at £1 each. I myself took up 20,000 shares for which I paid £20,000, and so had a shareholding that represented 14.2% of the company. We then set up a normal current account with an ordinary bank (not the merchant bank) and paid in our £140,000 (minus what we'd spent in setting up the company).

INCREASING THE SHARE CAPITAL

Before we continue the story of Chris's company, we need to know a little more about shares.

At a general meeting the shareholders can vote to authorize the creation of new shares to be sold to raise capital. If it is a public company, this is usually done via a finance company, which underwrites the new shares—that is, it guarantees that if the public doesn't buy them all, the finance company will take up the unsold shares at an agreed-upon minimum price. But if the public takes an optimistic view of the company, the outcome can be very satisfactory for the company.

If a private limited company raises new shares, it is usually by prior agreement with the shareholders for some known purpose, which means there is no doubt that they will be taken up.

> I have shares in a company that was formed to invest in technology. It created a large number of 1¢ nominal shares and went public on AIM. The new shares were underwritten at 5¢. As it happens, this was at the height of the dot.com craze when shares in companies trading in electronics went berserk. On the day they were issued to the public the quoted share price rose to 55¢ but felt to 27¢ a few days later. Within a few months the whole dot.com craze was shown to be idiotic, and the shares fell to 7¢. At least this particular company survived (unlike many other dot.coms) and its shares are now back to 12¢. Looking at these figures, which of these share prices—7¢, 12¢, and 55¢—could be thought reasonable?

Eventually, Chris's company became insolvent because of the failure of a subsidiary (see Chapter 4). It was rescued by being taken over, and this is how that worked: The nominal share capital of 140,000 £1 shares was increased to £1,400,000 by a new issue of 1,260,000 shares. The rescuer paid £1,260,000 for them in cash and that got the company out of being insolvent and gave him 90% of the company. I still owned my 20,000 original shares, but now they only represented 1.4% of the company.

GETTING CAPITAL FROM LOANS

Back to the company startup. On my rule of thumb that a contracting company needs £1 of capital for every £5 of turnover, that would have allowed us to aim at a £700,000 turnover, which wasn't enough to stay alive. So we persuaded the bank to give us a loan, which it did. That swelled the working capital we had available,

> Knowing that if you have bad luck you and your family will be put out into the street is a really nasty feeling!

and within a couple of years we had reached a turnover of £2 million.

The conditions attached to bank loans need a very careful look indeed. For our first loan the bank insisted that we, the executive directors, put up our family homes as security. So much for limited liability!

The main point, of course, is what interest rate is charged on the loan. But loan conditions also specify under what conditions the bank can call in the loan, its duration, and how it is to be paid off. Some are repaid in one lump sum at the end of a fixed period (when it is likely that a new loan will have to be agreed). Others are paid off at so much per year. The conditions in which the bank can cancel the loan need a particularly careful look: There have been some awful examples.

A company's relationship with its bank is most important. Chris, our CEO, made it his job to butter up our bank manager—a nasty little man, he was, but fortunately fond of golf, which proved useful. To be fair, though, the bank never let us down and at one time actually bailed us out in an emergency (see Chapter 8).

> Yet another grim example: A bank confused a successful one-man company with one whose name was similar, and called in a loan without which the poor chap couldn't go on trading and went bust. Then the bank discovered—whoops, sorry!—it was all a mistake! Sincerest apologies!
>
> By this time the man was ruined, his house had been repossessed, and his wife had left him. There was no way of restoring his business, even though the bank (generous as ever) offered to restore the loan. But in law, of course, the bank wasn't liable to do more than that—the loan conditions clearly entitled it to call the loan in any time it liked, and that's all the bank had done. Hard luck!

To sum up: The capital on which we set up our company was in part our own money, in part money from the merchant bank that took up shares and so shared our risk, but the largest sum of all was the loan from the normal bank. But why should a company want to rely on loans? Here's an example:

A company with $100,000 nominal capital has taken out no loans.

> The title "Bank Manager" hardly exists any longer: Normally the big banks now have staff in centralized offices whose job is to deal with the affairs of client companies over an area wider than that covered by one branch—but I daresay some of them are still quite keen on golf. And it is still a good scheme to do all you can to get them to feel happy about the company's stability.

Over the years it has retained $1.9 million out of its profits. In accountancy retained profit is reckoned to be the shareholders' property. Thus, together with the nominal capital and the share premium account, it makes up the total invested in the company by the shareholders: the *shareholders funds*.

Our company's capital is therefore $2 million. Thus, on paper each share represents a $20 investment by its owner. Let's suppose all this is available as working capital: $2 million can support a turnover of $10 million/year, and at (say) 5% profitability, the company makes a profit of $500,000/year at $5/year per share.

Suppose the company now takes out a $1 million bank loan at 6% interest, and passes the same sum on to the shareholders in cash, at $10 per share. The working capital remains unchanged at $2 million. Thus, the company goes on trading as if nothing had happened, making $500,000 profit as before. However, 6% interest on the loan must be paid, which reduces the net profit to $440,000 at $4.40 per share.

The shareholders have received $10 per share, which they can invest. If they do nothing more adventurous than put it into bonds yielding 6% (the same as the bank is charging on the loan), that brings in 60¢ per share. Thus, their total income is unchanged, but a 6% bond is less risky than the shareholding. Or they could always go for a more risky investment with a better return and increase their income. Either way they are better off.

The real objective of taking out the loan, however, is usually quite different. If the company is constrained by shortage of working capital, and uses the extra $1 million to boost its working capital to $3 million, it can now aim at a turnover of $15 million. And if it achieves that turnover at 5% profit, it will make $750,000 before deducting the interest on the loan. That means $6.90 per share after deducting the interest compared with the $5 per share, which the company would have earned without the loan—38% boost for the shareholders. No wonder they're happy.

Using loans to supplement investment in this way is a means of providing an advantage and, in an analogy to mechanics, it's called *leverage*.

This sort of logic means that companies are generally financed by a mix of invested capital and bank loans. The important thing is to get the balance between the two of them right. If a company borrows so much on loan that it is crippled by the interest payment, it will go under when bad times come. We've seen a lot of that in recent times.

SUMMARY

- **To start a business you first have to raise some capital from selling shares or getting loans.**
- **Contracting needs little fixed capital but depends on having enough working capital.**
- **Banks rely on fixed capital assets to secure loans, so contractors' borrowing power is limited.**
- **The amount of working capital limits the turnover that any company can sustain.**
- **Companies are financed most efficiently by a judicious mix of share capital, accumulated profit, and bank loans.**

Chapter 7

The Year's Business Plan

HOW THE BUSINESS WORKS

Now that the necessary capital is in place, we can start to trade, for which we must first plan what we hope to achieve in the year.

It's a good trick to ask people what they think is the most important objective in a business. Most of them will say "to make a profit," and they're wrong. The most important objective is *not to go bust*. It is quite easy to go bust while trading profitably (see Chapter 8), so making a profit is only the second most important objective.

> Business is like a jungle. In the real jungle every animal wants to procreate, and that is something you cannot achieve when you're dead. Survival is therefore the most important thing, which means that food (or flight) takes priority over fornication.

Before discussing how the system works, some definitions. These definitions aren't used by everyone, but the logic that follows depends on them, so if in doubt, refer back to them.

Turnover The year's output, measured by the total amount of sales invoiced (even if the money hasn't been received). It's commonly abbreviated as "T/O."

Direct Costs The costs that are directly due to carrying out the company's work, which will produce the turnover. The test whether a particular cost is direct or not is to ask whether the company would have incurred this particular cost if it hadn't executed this particular contract. Materials, engineering, site cabins, craneage, extra employees, and subcontracts arising out of the work are obviously direct costs, but on more doubtful details the company's accountancy system defines what is and what isn't a direct cost (see Chapter 13). It follows that direct costs can always be attributed to a specific contract.

Engineering Money: Financial Fundamentals for Engineers, by Richard Hill and George Solt
Copyright © 2010 John Wiley & Sons, Inc.

Overhead Costs that can't be attributed to any contract. They include such things as the accounts department or the receptionist's salary. Overhead is often called "indirect costs" or sometimes "fixed costs," based on the misleading assumption that the overhead stays constant whatever the turnover does. It does nothing of the kind. It's just that over a limited range of turnover it won't change very much.

Contribution What's left of the contract price of a particular contract after the direct costs of that contract have been paid. The contribution on all the contracts goes into the company's kitty to pay for the overhead, and what's left is profit.

Profit What's left at the end of the fiscal year, when the company has paid all its direct costs and overhead.

So, taking the totals in a year's trading

$$\text{Turnover} - \text{Direct Costs} - \text{Overhead} = \text{Profit}$$

Obviously, this only works on the basis of the definitions set out above.

PLANNING FOR PROFIT

Although survival is more important than profit, we do, of course, want to make as much profit as we can. It is clear that a single contract on its own doesn't make a profit. However good some contracts might look individually, profit depends on the sum of the contributions from all the year's contracts being greater than the overhead.

Normally, then, a higher turnover yields a greater profit, and we plan for the highest turnover. The first thing is, therefore, to see how much that is. It depends on three possible limiting factors: how much work the company can actually perform, how much the working capital will sustain, and how much of the market we can command. The smallest of these three will show the most for which to plan.

Next, we must estimate what our overhead will be at the lowest of these limitations and what percentage profit our operation can reasonably expect. Profit levels vary enormously both between and within different industries. For contracting they are generally low, with the exception of some specialized trades that need unusual skills or expensive and costly equipment.

The reason is that contracting needs little fixed capital and that means it is relatively easy to start a contracting company. If that were not the case, we would never have been able to raise enough to get our company started. But others could (and did) do the same as we had

At one of our earliest board meetings, Chris said he wanted us to make 10% profit on turnover. It's not done, I said, no one in this trade makes more than 5%. We tried but, sadly, I was right—the most we ever achieved, just for one year, was 6%.

done. New companies spring up like mushrooms, and competition from them is always ferocious. It is said that contracting is the last truly competitive industry there is.

Once these numbers have been estimated, they fix the total contribution that the year's work has to bring in, and therefore the average contribution on all contracts. The result might work out like this:

Turnover (T/O)	$40 million	That's what you think you can do given the company's capability, its working capital, and the market conditions.
Direct Costs	$30 million	That depends on the type of work you are doing and also on your accountancy system.
Overhead	$7.5 million	That estimate should match the turnover of $40 million. It would leave you.
Profit	$2.5 million	Which is 5% on T/O—the most that you thought the market would stand.

To meet this plan, therefore, you need to generate a total contribution of $10 million from all your contracts in order to cover overhead and profit. (This total contribution is usually called the *gross profit*, but as it has nothing to do with profit it is a rather confusing term, and we will avoid it except when talking about fiduciary accounts.)

That's 25% of the turnover, so on the face of it each contract for the year must aim to make the average contribution of 25% of its contract price. In practice, some will do better and some worse, but the 25% target will serve as a guide when pricing contracts.

The graph in Figure 7.1 shows how the company's performance varies with the turnover. It is based on the simplifying assumption that all the contracts carry a 25% contribution, so the contribution increases linearly with turnover. At point A, when the turnover is $30 million, the contribution is $7.5 million and just covers the overhead without leaving anything for profit. This is called the *breakeven point*. Point B is the planned performance, which we calculated above, with a turnover of $40 million and a contribution of $10 million, giving a profit of $2.5 million after the overhead has been paid.

Figure 7.1 also shows that, because profit is a small difference between large numbers, minor changes in the main numbers make a big difference to the profit. Point C shows that exceeding the plan by 25% (a turnover of $50 million) promises twice the profit! But, of course, that isn't true: It makes two of our basic assumptions false. First, overhead doesn't remain constant with changing turnover. This is why it is so misleading to use the term fixed costs when, in practice, a significant increase in turnover would certainly lead to an increase in overhead.

The second false assumption is that we worked out that the $40 million turnover was the maximum that the company can achieve, because of the market, the limitations of the company's staff and equipment, or its working

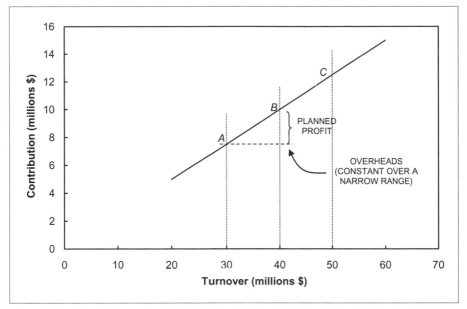

Figure 7.1 Breakeven and profit.

capital, which, at 5% of turnover, would be $20 million. Engineers in particular know that overloading the workforce will lead to costly mistakes and delays, but are rather liable to ignore the limits imposed by capital, which leads them into problems.

OVERTRADING

Any company needs working capital to get by. Buying materials, employing subcontractors, and the like all make demands on it. Then there are the constant outgoings of salaries, rent, and so forth. The company's contribution is supposed to bring in enough to cover all these, but there is always a timelag between money going out and money coming in. That means that the greater the turnover, the greater the deficit that this causes. Big irregular bills are particularly difficult to meet—the commonest event that brings companies down is the annual tax bill. Whatever the cause, if there's nothing in the bank to pay the bills, the company is insolvent. If this has happened because the company undertakes a bigger turnover than its capital will support, it is called *overtrading*.

The people who are most likely to fall into this trap are brilliant engineers who have a good idea and set up their own company to exploit it. Their idea is so good that the company is profitable and grows. But, if they don't realize that they must increase the company's working capital to keep up with the turnover, sooner or later it will fail, even if the company is trading at a profit. Some rich old company can then buy it up, pay the outstanding bills, and cash in on the success.

One of my company failures came about like that, partly because our sales executive fell into the error of thinking of overhead as fixed costs—that's why I dislike the term so much.

That assumption led him to argue that once he had sold enough contracts to meet the planned turnover, we would have paid off the overhead. The contribution from any further sales would then be pure profit. So once he had got enough orders to fulfill our year's plan, he went on making more offers at a reduced contribution. That meant he was undercutting the competition's prices, and made it easy for him to make lots more lovely sales.

But overhead isn't fixed, and an increase in turnover will make it rise—it's just that over a narrow range the change may be relatively small. So the argument is flawed from the start. He routinely exceeded the year's sales target by 10% or more. I warned him that we would run out of cash, but he pooh-poohed that. Our biggest single direct cost on contracts was for steel pressure vessels, and he reckoned that on the terms of payment that we had on our contracts, we would get paid by our clients before we had to pay the steel fabricator for the vessels. In practice, it didn't work that way. Besides, the engineering department was so swamped with work that many contracts suffered expensive errors and delays. After 5 years of this regime we went bust and had to sell the company. So much for exceeding the plan.

Some people have told me always to distrust sales executives on principle. Like the culprit in this example, they concentrate their minds so hard on the business of getting contracts that they ignore little problems such as survival and making a profit. This man's predecessor in our company, on the other hand, was really good at his job. He kept saying that any fool can sell at the lowest price, but the chap who can sell at a high price is the one who is really good at his job. It's a philosophy I strongly recommend to everyone.

MARGINAL SELLING

But what if the sales don't come up to the target? The graph shows that the result is to make the profit fall sharply. Again, if the sales are below the target, the overhead will probably fall a bit—phone bills will be smaller, employees who leave won't be replaced immediately, and so on. But the decrease won't be enough to maintain the profit.

The plan was based on filling the company's capability to do work and the working capital available to finance it: It's the market that has failed its promise. What's to be done?

These are the conditions in which one is right to sell contracts with a reduced contribution. In plain English, cut-price selling makes it easier to get contracts. The logic here is the exact opposite to that proposed by my former colleague, the keen sales exeutive. If the company has underused capacity, then any contribution that can be made from it is better than none.

Here's the logic: Suppose you plan a journey alone in your automobile, and a chap says "give me a lift and I'll pay my share of the fuel." That won't even cover half the real cost of the drive, but taking him actually costs you nothing. If you turn him down, you get nothing at all. Thus, you're better off if you take him and his money. Spare capacity is the key.

Like many process plant contractors, we had erection teams to get the equipment erected and installed and commissioning engineers to get it operating well enough to pass its acceptance test. These are highly skilled jobs—once you've got good people you don't let them go.

Our work schedule occasionally had no contract ready for them to work on. Then we would offer their services to other companies at what were (on paper) uneconomical rates. The cut price meant we could usually procure work for them and at least bring in some contribution. If they had remained idle, that would have brought in none.

SUMMARY

- **A company needs a plan for the year in order to achieve its maximum potential.**
- **That plan has to set targets for turnover, overhead, and profit.**
- **If it's a good plan, then any serious change from it can't be good.**
- **If the actual turnover significantly exceeds that planned, the company may get into trouble.**
- **If it looks unlikely that the target turnover will be reached, marginal selling (by cutting the contribution on contracts) can be a useful partial remedy.**

Chapter 8

How Not to Go Bust

NEED FOR WORKING CAPITAL

There is only one way in which companies fail—by not having money with which to pay their bills. That happens if their working capital has run out: whether they are making a profit or not has little directly to do with that.

The amount of working capital that companies need, measured as a percentage of the turnover, varies a lot between different types of industry and even between different companies in the same industry. It seems as if each company's circumstances dictate how much is needed, so that the figure for any individual company tends to remain fairly constant, however much the turnover rises or falls. Typically, contracting companies might need working capital equivalent to a fifth of their turnover. Whatever the percentage, it is the amount that gives the company enough flexibility to survive normal trading fluctuations.

It's important to understand that turnover is not the most important thing for the company. If that turnover isn't profitable, there's going to be a problem. I once worked for a CEO who had been a salesman. Before a tender was submitted, the managers would have a meeting to discuss and agree on the price that we would offer. The price makeup sheet had, as part of the contribution to company overhead, a notional profit of 2.5%. Given the opportunity, our CEO would try to see the client after submission of the price and would almost invariably offer a 10% discount. In vain I tried to point out that this meant that most of the contracts we won would make a 7.5% loss (unless the contract manager was clever enough to find a way of making savings). He was happy that the company's turnover was increasing year by year. Sadly the company went bust, but with a full order book.

Here are some cases to show how this works, based on the $40 million turnover company with 20% working capital of $8 million, which we have previously used as an example.

Engineering Money: Financial Fundamentals for Engineers, by Richard Hill and George Solt
Copyright © 2010 John Wiley & Sons, Inc.

CASE 1

In one year, the company has a loss of 2% on turnover—that is, a loss of $800,000. This reduces its working capital by 10% to $7.2 million. That will not make a massive difference to next year's trading when, with luck, business will be better, the loss recouped, and the working capital restored. A modest and occasional loss doesn't bring a healthy company down.

My wife inherited shares in a family company making expensive cloth at an ancient factory in a small town in Somerset. Around 1900 it was very profitable, but by 1950 Somerset had become the wrong location for this trade. The family members were kindly people: They never sacked anyone because there was no other employment to be had locally. For years the company's annual audit showed a loss, which it survived because it owned a lot of land locally (fixed capital) that could be sold to make more money available for working capital. In the (inevitable) end it ran out of saleable bits of Somerset and become insolvent.

CASE 2

The company has a major contract worth $12 million, and there is a problem that delays a part payment of $4 million. The company still has $4 million of working capital with which to carry on, and, if the delay is not too long, it ought to survive. The most obvious action is to go to the bank for extra finance, such as a temporary increase in its overdraft limit, especially if the cash crisis is expected to be short term.

CASE 3

The company is working on two contracts each worth $12 million. Both of these suffer some setback: A payment of $4 million is delayed on each, which reduces the working capital to zero. Unless cash can somehow be raised quickly, the company cannot pay its bills and is insolvent. This can happen as easily to a company that is making a profit as one running at a loss, but in practice there is a big difference. A profitable company is more likely to command the bank's confidence and to be able to persuade it to help the company through its difficulties.

This shows that companies must not take on contracts so big that a serious setback can break the company. In the fatal case 3, the troublesome contracts are indeed only 30% of the turnover each, but with two of them in trouble that's enough to bring the company down.

Large companies, for example, the petrochemical giants, have lists of potential contractors who are considered fit to bid for jobs. These lists show the maximum value of contracts for which they should be considered. Contractors who have grown since getting on to one of these lists have quite a struggle to get promoted into a bigger league.

Our company avoided tendering for contracts so big that they could sink us, and this works both ways—clients don't want to employ contractors who might go bust while halfway through the job.

What's to be done, though, by a contractor threatened with being awarded a contract so big that he shouldn't really take it on? For example, suppose that while a contractor is negotiating a contract near his size limit, the client increases the size of the project. If the contractor lands the contract, he runs two risks: First, the company can't cope physically—not enough manpower, machinery, and the like—and second, it hasn't got the working capital.

The first danger is obvious enough, and so is the action needed to overcome it. Unfortunately, the money problem is easy to ignore—especially by engineers who aren't supposed to know about money. At a moment when everyone is licking their chops at the prospect of getting a great chunk of work, it's easy for them to think, "Oh, don't worry, it'll be OK."

The best planned jobs can come unstuck. There's always a risk of trouble and delay by things such as weather, strikes, and accidents. The conditions of the contract may protect the contractor against *force majeure* (see Chapter 11), but, even then, the delay in getting paid in a small contract can be a nuisance. With a very big contract, it can break the company. Of course, small and big in this context are relative to the size of the contractor's turnover.

> Our company had a big contract for a Cuban sugar mill, at the height of the Cold War when communist Cuba was practically cut off from the western world. We were to be paid CIF (see Chapter 18). We hadn't realized that at that time there were only three boats a year to Cuba, and we just missed one. The entire consignment, all crated up, spent months sitting at the docks. We'd had the cost of producing it, and hadn't been paid a penny. Result, a cash flow crisis.
>
> The bank was shown the detail, so it knew that we would get paid within a few months. Also, we'd always taken a lot of trouble to ensure they knew (or thought they knew) what a splendid and reliable outfit we were. So they increased our overdraft limit—at their usual high interest rate, of course—and when the next ship came in we were out of trouble.

That kind of disaster can happen as easily to a company that is making a profit as one running at a loss.

SUMMARY

- **Companies fail if they run out of working capital.**
- **This is true even if they are making a profit.**
- **Small, temporary losses are not usually fatal.**
- **A company must not take on contracts that are so large that they could cause it to fail.**

Chapter 9

Cash Flow

WHAT'S CASH FLOW?

It boils down to this: To stay alive, a company must always have enough money to pay its bills. It must, therefore, regulate the flow of money in and out to reduce the risk of running out of cash. Let's consider the tank analogy introduced in Appendix 1. It's all about controlling the flows into and out of the tank to ensure that there's always something in the tank. Control of cash flow is one of the most important aspects of running any commercial venture and, in the contracting business, it's the job of the contract manager.

The monthly bank statement of your personal current account shows how your cash flow has gone, with sums of money coming in, payments going out, and the daily balance during the month. A company's bank statement does the same.

This is important in all forms of commerce and industry, but far more so in contracting. That's because, in a contracting company, the outgoings are of two kinds—a fairly steady flow of the regular costs in running the company—paying salaries, rent, and so on. Most of the money, however, goes out irregularly and sometimes in very large chunks—these are for the materials and services needed to carry out its work. Payment coming in is similar in that it tends to be in irregular, large sums, received for work as it is completed. The monthly balance goes up and down and can be shown as a table or a graph. Let's look at an example to see how this works. Suppose we have a contract valued at $5.5 million with the price made of the elements shown in Table 9.1 (see Chapter 14):

Table 9.1 rather nicely demonstrates the curse of the computer (see Chapter 24): The estimates are all given to the nearest $100 (and that is probably more accurate than the facts justify). When the percentage additions are made, the spreadsheet calculates them to the nearest $1, and that is how the price is presented which is, of course, absurd.

Now let's assume, initially, that we get paid in a single payment on completion of the contract. The contract is completed in 12 months. Thus, allowing

Engineering Money: Financial Fundamentals for Engineers, by Richard Hill and George Solt
Copyright © 2010 John Wiley & Sons, Inc.

Table 9.1 Contract Price Makeup Sheet

Item	Quantity	Rate	Amount	Total	
Materials-mechanical			$950,000		
Materials-electrical			$500,000		
Materials-control & instrumentation			$200,000		
Materials deliveries			$20,000		
Sub Total Materials				**$1,670,000**	Note 1
Mechanical installation sub contract			$100,000		
Electrical installation subcontract			$60,000		
Sub Total Installation				**$160,000**	Note 2
Process engineer	300 hrs @	$120.00 /h	$36,000		
Mechanical engineer	650 hrs @	$100.00 /h	$65,000		
Electrical engineer	450 hrs @	$100.00 /h	$45,000		
Controls engineer	400 hrs @	$120.00 /h	$48,000		
Drafting	500 hrs @	$80.00 /h	$40,000		
Sub Total Engineering				**$234,000**	Note 3
Commissioning	1200 hrs @	$100.00 /h	$120,000		
Sub Total Commissioning				**$120,000**	Note 4
Conract Manager	2000 hrs @	$120.00 /h	$240,000		
Procurement	500 hrs @	$80.00 /h	$40,000		
QA	300 hrs @	$100.00 /h	$30,000		
Safety	400 hrs @	$100.00 /h	$40,000		
Sub Total Management				**$350,000**	Note 5
Site offices	50 weeks @	$800.00 /week	$40,000		
Craneage	200 days @	$800.00 /day	$160,000		
Support staff	8000 hrs @	$60.00 /h	$480,000		
Telephone			$20,000		
Stationery/printing			$20,000		
Transport			$20,000		
Sub Total Site facilities				**$740,000**	Note 6
Interest	2.0% of Contract Price	$110,050			Note 7
Agents' fees	1.0% of Contract Price	$55,025			Note 8
Fixed price	1.0% of Contract Price	$55,025			Note 9
Licence fees	0.5% of Contract Price	$27,513			Note 10
Insurances	1.0% of Contract Price	$55,025			Note 11
Contingency	5.0% of Contract Price	$275,126			Note 12
Sub Total other costs				**$577,765**	
Total Direct Costs				**$3,851,765**	Note 13
Contribution	**30.0% of Contract Price**			**$1,650,756**	Note 14
CONTRACT PRICE				**$5,502,521**	Note 15

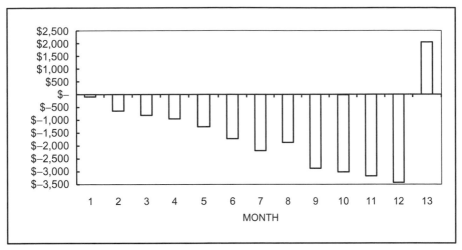

Figure 9.2   Bar graph of contract cash flow.

for the month's delay between issuing the invoice and receiving the check, we get paid in full in month 13. The monthly contract cash flow table is shown in Figure 9.1.

For simplicity the interest for each month is calculated on the previous month's overdraft. That's not exactly right because the bank calculates interest on a daily basis, but it makes the spreadsheet easier. We can show this as the bar graph in Figure 9.2.

The money left in the bank at the end of the contract is the contribution plus contingency, in this case $1,581,085 or 29% of the contract price. That's not quite as much as we planned for in the price makeup sheet.

WHAT DOES THIS TELL US?

The two important things to look for in Figure 9.2 are: how great are the peak variations (plus and minus) from the average balance over a long period, and what is the average balance over a long period.

Looking first at the variations, if the account is in credit most of the time, it suggests that the company is not using its working capital to the best advantage. On the face of it the company has cash lying around that is not doing any real work. If the company can't use it, the shareholders should get it as dividend.

Having said that, if a company knows it is going to have spare cash just for a few months, it should at least invest the surplus on a short-term basis and get the benefit of the interest that that can bring in.

If the minus variations are regularly near the overdraft limit, the company is living dangerously. Either it isn't managing its cash flow very well or it needs more capital, or both.

MONTH	1	2	3	4	5	6	7	8	9	10	11	12	13	Total
Opening balance	0	-87,667	-643,778	-810,810	-949,983	-1,257,316	-1,720,961	-1,720,961	-1,870,469	-2,872,222	-3,031,324	-3,175,752	-3,425,383	
Payment received													5,502,521	5,502,521
Total receipts	0	0	0	0	0	0	0	0	0	0	0	0	5,502,521	5,502,521
Site establishment			74,000	74,000	74,000	74,000	74,000	74,000	74,000	74,000	74,000	74,000		740,000
Project management	29,167	29,167	29,167	29,167	29,167	29,167	29,167	29,167	29,167	29,167	29,167	29,167		350,000
Engineering	58,500	58,500	58,500	29,250	29,250									234,000
Equipment supply					167,000	334,000	334,000		835,000					1,670,000
Installation						16,000	16,000	32,000	48,000	32,000	16,000			160,000
Commissioning												120,000		120,000
Miscellaneous		467,714												467,714
Interest at 10% per annum	731	5,365	6,757	7,917	10,478	14,341	14,341	15,587	23,935	25,261	26,465	28,545		179,722
Total costs	87,667	556,112	167,031	139,173	307,333	463,644	467,508	149,508	1,001,754	159,102	144,428	249,631	28,545	3,921,436
Closing balance	-87,667	-643,778	-810,810	-949,983	-1,257,316	-1,720,961	-2,188,469	-1,870,469	-2,872,222	-3,031,324	-3,175,752	-3,425,383	2,048,593	1,581,085

Figure 9.1 Contract cash flow.

As to the average balance, the important thing is that the datum is not necessarily fixed. If a company is heavily into overdraft most of the time, and the bank can be persuaded into giving it a loan, that would shift the datum. If, having done that, the bank doesn't reduce the overdraft limit, it shifts the average value and boosts the working capital. Interest on loans is lower than on overdraft, so it is better to finance a company by loans than by overdraft.

Our company had a period when we had a lot of cash in the bank. We put the surplus into local authority loans for a few months and the interest we got from that accounted for a quarter of that year's profit. (That's not surprising if you remember that in contracting the profitability—that is, the profit as a percentage of turnover—tends to be very thin). But it didn't take long before things reverted to normal and we were in overdraft again.

In practice, though, most companies are also in overdraft most of the time. The logic is quite simple. In Chapter 7 we had a modestly successful contractor making a profit that is 25% of the working capital. As that return is far greater than any overdraft interest rate, it's sensible to use as much of the available overdraft as is safe to boost the working capital. The danger is that it reduces the safety margin that the overdraft facility gives, so it is important not to push it too far.

It's also important to avoid violent variations—as far as that is possible. A high credit balance for a short period feels nice but is useless. Large overdrafts are dangerous if they get near or beyond the limit set by the bank. Management of the cash flow should aim at getting the incomings and outgoings to balance one another, thus keeping the swings to a minimum.

Better still, of course, is to get all payments to come in as early as possible and pay out as late as possible.

PROGRESS PAYMENTS

Any contract for a project sets down when and how much the client will pay the contractor (see Chapter 18). If you have a plumber do some minor work in your house, you expect to pay the full amount when the job is finished. Commission a complete new kitchen, and you will have to pay a sizable fraction of the contract price before the kitchen contractor starts the job—that's because he in turn has to order and pay for the new appliances.

The bigger the project, the more important this becomes. Obviously, if a contractor, with a working capital one-fifth of his turnover, had to stand the full cost of completing his projects before he has any money coming in, he would run out of cash within a few months. There is an obvious conflict of interest here: The contractor wants to get paid as early as possible, while his client (who is also minding his cash flow) wants to delay shelling out as long as he can. What payments are to be made on a contract, and when, are

MONTH	1	2	3	4	5	6	7	8	9	10	11	12	13	Total
Opening balance	0	−87,667	−93,526	−255,972	−390,522	407,311	−45,855	−45,855	1,194,226	208,059	1,448,523	1,329,356	1,656,442	
Payment received	0	550,252	0	0	1,100,504	0	0	1,375,630	0	1,375,630	0	550,252	550,252	5,502,521
Total receipts	0	550,252	0	0	1,100,504	0	0	1,375,630	0	1,375,630	0	550,252	550,252	5,502,521
Site establishment			74,000	74,000	74,000	74,000	74,000	74,000	74,000	74,000	74,000	74,000		740,000
Project management	29,167	29,167	29,167	29,167	29,167	29,167	29,167	29,167	29,167	29,167	29,167	29,167		350,000
Engineering	58,500	58,500	58,500	29,250	29,250									234,000
Equipment supply					167,000	334,000	334,000		835,000					1,670,000
Installation						16,000	16,000	32,000	48,000	32,000	16,000			160,000
Commissioning												120,000		120,000
Miscellaneous		467,714												467,714
Interest at 10% per annum		731	779	2,133	3,254	0	382	382	0	0	0	0	0	7,662
Total costs	87,667	556,112	162,446	134,550	302,671	453,167	453,549	135,549	986,167	135,167	119,167	223,167	0	3,749,376
Closing balance	−87,667	−93,526	−255,972	−390,522	407,311	−45,855	−499,404	1,194,226	208,059	1,448,523	1,329,356	1,656,442	2,206,694	1,753,145

Figure 9.3 Contract cash flow with stage payments.

therefore a vital part of the original contract agreement. We'll get back to them later (see Chapter 22).

Once the terms of payment are fixed by the contract, and the points at which payment will be made become clear, one of the contractor's main aim is to reach them as quickly as possible. That needs good planning (see Chapter 21) and can often mean spending extra money. The whole company, and particularly the engineers and the management accountants, must work closely together to make the best of this task.

Let's assume that our contract has terms of payment that give us stage payments as follows:

On larger jobs we often had a contract that gave us a first payment on delivery of the engineering drawings or other documentation. As soon as the order was confirmed, therefore, the engineering department began to work flat out, possibly putting people on overtime or engaging freelance draftspeople to speed up the work. When the complete drawings were ready, they would be sent by e-mail to save time. Of course, there's no need for the contractor to have the complete set of drawings ready at the start of a contract, and the additional effort to get them done costs extra. But getting paid a large sum early makes it all worthwhile.

10%	With order
20%	On approval of drawings
50%	On delivery of all equipment
10%	On completion of installation
10%	On completion of commissioning

That gives a completely different project cash flow, as Figure 9.3 shows.

Now the graph looks like Figure 9.4. Better still, the contribution has increased to $1,753,145 or 32%, which is better than we anticipated. And that

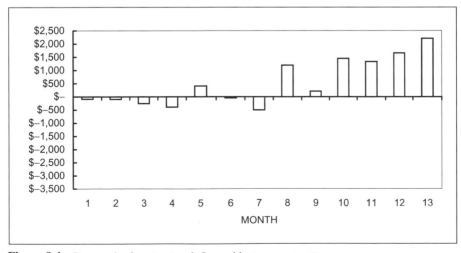

Figure 9.4   Bar graph of contract cash flow with stage payments.

calculation doesn't take into consideration the fact that during the periods when we have cash in the bank it can earn us some interest.

So cash flow and, in particular, terms of payment, is not only important in terms of the company's survival, it also affects its profitability. It's part of the contract manager's job to control that cash flow.

RETENTIONS

Sometimes terms of payment include a *retention*, which is a part of the contract price (usually 5 or 10%) that the purchaser retains (i.e., he doesn't pay it to the contractor) for a period of 6 or 12 months after completion of the contract. The contractor gets his money after this time only if the purchaser is satisfied with the job. The theory is that, if the contractor hasn't been paid in full, he will take his postcontract responsibilities more seriously and carry out any rectification work and make good any latent defects.

The original concept was that the retention would represent the contractor's profit and this would be an incentive to perform. These days, of course, with profit margins greatly reduced, a 10% retention represents a significant part of the contribution on a contract, and waiting for it for 12 months can be a serious cash flow problem. At the very least, the contractor will probably need to borrow money to cover the retention, and this will increase the interest charges.

PAYING LATE

The best way of improving cash flow by paying as late as possible is by good planning and skilled purchasing. Goods and services should arrive when wanted and not before, so that the invoices for them only become due at the latest possible time in the project's program.

The normal rule for invoices is that payment is due one month after the date of the invoice. Sometimes it's a calendar month, sometimes 30 days, 45 days, or even 60 days. Paying invoices later than that can be done legally or illegally. One legal way is for the contractor to agree on some longer period of credit with his suppliers—this is called *supplier credit*. A civil engineering contractor, for example, is likely to need a lot of concrete. He may agree with one supplier of ready-mix that he will not buy the stuff from any of his competitors: in exchange for that the supplier agrees to give him 2, or even 3 months before he needs to settle invoices.

The illegal way is all-too common: the company simply doesn't pay its bills on time. Many smallish businesses have at one time or another been so short of cash that they deliberately sit on their invoices even though they are due. The CFO confers with the CEO to decide which of the bills they really have to pay now—because failure to pay would upset an important supplier, for example—and which creditors can safely be left to wait for their money.

Occasional events like this would be tolerable, but in recent times late payment has become endemic particularly in the UK. There the rot started in the 1970s with large private and even some state-owned industries making late payment their policy. The Central Electrical Generating Board, for example, was a nationalized company that ran all UK power stations: For some types of equipment they were almost monopoly customers and could afford to ill-treat their suppliers with impunity. When these big clients decided to pay late, their suppliers ran short of cash and couldn't pay their own bills, and so on down the chain. No one dared to take their clients to court for their money for fear of offending a vital customer, and anyway the process of recovering the debt would have taken so long and cost so much that it would have been commercial suicide. Although European legislation now imposes quite heavy interest charges on unpaid invoices, its effect has been negligible for the same reasons.

Late payment became a national disease in the UK, and a lot of small companies went to the wall because of the trickle-down effect— not through mismanagement on their part, but because they were helpless victims of the big organizations' refusal to pay on time. Internationally, UK companies earned themselves a reputation for being bad payers.

A Chinese student who had worked in a Chinese restaurant in New York City told me that this restaurant had acquired a reputation for being rotten payers. When its suppliers got fed up with this, they simply sent the restaurant rotten food.

Readers of P. G. Wodehouse novels will know that in his time the young gentlemen at Oxford and Cambridge would run up tailors' bills for their whole 3- or 4-year stay, which they weren't expected to pay before they'd inherited the family fortune.

At that time traditional industries were financially so profitable that cash flow wasn't very important. In 1968 I worked on a contract we undertook jointly with a Dutch company. Some months after the contract was completed and the books closed, we were embarrassed to receive an invoice for painting—the Dutch subcontractor was still in the habit of only sending out his invoices at the end of the year.

What means the debt factors use to collect the money I don't know. But I was once asked to call on a client's offices that happened to be on my way home. They owed us a large sum and our accountant had been told "the check is on the CEO's desk, sir"—clearly it had been put on the "let 'em-wait" pile. "Can I make myself as disagreeable as I like?" I asked, "yes" was the answer.

It was a Friday evening. Their office was a small high-rise block, with a basement parking lot. I parked my car blocking the ramp down to it, and went up to the accounts office. "I've come to collect our check" I said, "which we're told is on your CEO's desk. And, by the

In desperation, a company can get paid most of the money it is owed by "factoring out" the debts to professional debt collectors. This maintains the cash flow but loses a crucial percentage of the total sum.

way, I seem to have parked across the exit to your parking lot, so until I've got it, nobody is going to be able to go home."

It was perfectly true. The check must have been on the CEO's desk because I drove off with it in my pocket a few minutes later.

SUMMARY

- Contracts and companies must be run in such a way that they don't run out of cash.
- This means predicting and managing the flow of money in and out.
- On contracts of any serious size, there is usually some agreement for progress payments by the purchaser as work proceeds.
- The amount and timing of these progress payments are crucial.
- As well as the danger of running out of cash, not getting paid for work means having to borrow more, and that means increased interest payments that slash profits.
- Illegal late payment causes serious problems.

Chapter 10

What's a Contract?

AN APOLOGY

Sorry, but some of the stuff in the next few chapters is "legalese" and can become quite boring unless you've got the sort of mind that soaks up the odd bit of Latin. But bear with me, it's worth it—you'll be able to baffle the opposition!

Lawyers and medics use Latin, which their clients do not understand, and this gives them an air of superiority. This is why they can charge higher fees than we engineers who use easily understandable English.

Legal, of course, refers to the law. Laws are rules made and written down by governments as statutes (statute law) or developed by judges through decisions made in courts and similar tribunals based on precedent (common law). Common law is mainly derived from Anglo-Saxon law and forms the basis of the English legal system. Precedent refers to decisions taken previously in similar cases in any Anglo-Saxon jurisdiction. As a result of the influence of the British Empire, Anglo-Saxon systems are found in the United States, Canada, Australia, New Zealand, Singapore, Hong Kong, Malaysia, India, and other former colonies, and this type of law is generally the basis for international contracts. Contract law in these countries has many similarities, but details vary from country to country and even within countries—Scottish law is different from English law, but both are subservient to European Union law—and some legal systems are quite different. It's very important, therefore, to know which country's law governs a particular contract.

It's all a bit complicated, but engineering contracting is an international business—a Japanese contractor may be building an oil refinery in China for an American oil company—so the contractor needs to know which law governs the contract. Some religions have their own codes such as Jewish Halakha and Islamic Sharia. The latter is becoming increasingly significant as trade with Islamic countries increases.

A contract is the starting point for virtually all engineering projects. It is an essential part of the project, and without a contract no major engineering project can be undertaken.

Engineering Money: Financial Fundamentals for Engineers, by Richard Hill and George Solt
Copyright © 2010 John Wiley & Sons, Inc.

IT'S AN AGREEMENT

An *agreement* occurs when the parties discussing or negotiating an arrangement between them have reached an identical conclusion—they are *ad idem* (I warned you about the Latin). We make agreements with each other all the time: I may agree to meet a friend for a drink at a bar. However, if I fail to turn up, neither of us would expect the friend to be entitled to sue me. A contract is an agreement between two (or more) parties that is enforceable in law provided that certain conditions are observed. This means that one party will be liable to pay compensation to the other party if it breaks the agreement, and a court (or other judicial tribunal) can make this happen.

As far as we are concerned as engineers, a contract is almost always a promise by one party (the *contractor*) to supply certain goods (e.g., equipment such as pumps and valves) and/or services (e.g., design, construction, or installation) in return for payment of money by the other party (*the purchaser* or *the client*—we'll use these two terms interchangeably). For a formal contract to be valid, the following elements are necessary:

- There must have been communications between the parties showing their intention to deal with one another—these are called *offer* and *acceptance*.
- The parties must be legally capable of contracting (e.g., companies must be incorporated).
- There must be a *consideration* or a *quid pro quo*, which means that the purchaser must promise to remunerate the contractor for executing the contract.
- There must be a written form of contract. (This is not strictly true for all types of contract but, because engineering contracts are complicated and it is very difficult to get reliable evidence about verbal agreements, it is the general rule as far as we are concerned.)

  > As Samuel Goldwyn, the famous film producer, once said "A verbal contract isn't worth the paper it's written on."

- The contract must be legal, that is, the objective of the contract must not be forbidden by law. (You can't employ a contractor to rob a bank or to manufacture illegal drugs.)

  > If you employed a contract assassin to kill your boss and he failed to do it, you couldn't sue him for breach of contract—the contract would be invalid because the objective was illegal. You might find yourself in all sorts of other trouble, too.

- It must not be impossible to achieve the object of the contract. (The contract becomes void if, for example, something comes to light in the course of the contract that makes it impossible to carry out certain work, and that neither party was aware of at the time of contracting.)

Contracts may be rendered impossible by events such as floods or war. This is called *force majeure* (see Chapter 11).

- The consent of the contracting parties must be genuine, that is, not marred by mistake, misrepresentation, or fraud.

For an agreement to exist between the two parties, there must be an understanding of what has been agreed to and this understanding must be demonstrable. This is the purpose of the *contract documents*. These are usually presented in the form of a tender document, which is issued by the purchaser to set out exactly what goods and/or services he or she wishes the contractor to supply (the technical section) and the terms and conditions that he or she wishes to impose on the contractor (the commercial section). In practice, this is not always as easy as it sounds.

HOW PROJECTS HAPPEN

It's quite hard to define a *project* properly, but for our purposes it is some sizable undertaking to build or produce something durable. It might be a road bridge, or a sugar mill, or a reservoir, or an oil pipeline. The project is commissioned by the purchaser and executed by the contractor.

There is a simple logical sequence for bringing this about:

- The purchaser perceives a need for some new or additional facility.
- He defines in general what that need is.
- He makes a rough estimate of how much this might cost and uses an evaluation tool like such as net present value (NPV) or simple payback to decide whether the benefit makes the expenditure worthwhile (see Appendix 3).
- He decides that he should commission a project to satisfy this need.
- He commissions a detailed specification of the project (he may employ a consultant to do this or he may do it himself) that would satisfy the need.
- He makes a better estimate of the probable cost, sets a budget to meet it, and ensures that he has the means to pay.
- He alerts potential contractors to the prospect of carrying out the work and invites them to discuss details with him.
- He issues a definitive specification of the work to be done and gives an indication of the terms of the contract he proposes to let.
- He discusses the specifications with potential contractors to iron out details and possibly receive some suggestions from them.

- The contractor(s) prepare offers to do the work and submit them to the purchaser.
- The purchaser considers the alternative offers and selects the one that (if any) appears best (there may be a short-list stage first).
- Purchaser and contractor(s) meet to consider and agree-upon details.
- The purchaser awards the contract to the selected contractor, who agrees to perform it, and the contract is signed.
- The contractor performs the work set out in the specification.
- Any changes from the specified work should, if possible, be agreed upon while the work proceeds.
- The purchaser pays the contractor as the work reaches points at which payment is due.
- When the work is completed, the purchaser checks that the result meets the specification.
- Purchaser and contractor are happy with the outcome.

If only real life were like that!

In reality, there are always some projects that are allowed to proceed without some of these jobs having been performed properly or at all, and that usually means trouble.

TENDER DOCUMENTS

Tender documents vary enormously depending on the industry, the type and size of the contract, whether it is a greenfield project or modifications to an existing plant, and who has prepared them. Tender documents are often prepared by civil engineers and concentrate on general specifications and priced bills to allow *remeasurement* (remeasurement is done by quantity surveyors). In the chemical industry the documents are more often prepared by process engineers who deal more in terms of process performance and lump-sum prices.

Tender documents prepared by consultants frequently contain large amounts of irrelevant material designed to "pad out" the documents—a thick document looks like good value for the consultant's fees. While it is difficult to generalize,

> The thickness of tender documents used to be used as a quick guide to the contract price, at about $10,000/mm but, over recent years, this has proved unreliable.

most tender documents will contain at least some of the following sections (in no particular order):

Scope of Work: This should define the work to be executed under the contract and should provide background details such as why the work is to be done and what impact other associated contracts might have on the work.

Particular Specification: This section (usually the most important for the proposals engineer) sets out specifications for processes and equipment to be used on the project. It may consist of a simple performance specification (e.g., guaranteed quantity and quality of product); a *design and build* specification (see Chapter 17), which details the process route to be used but not the detailed design; or a *detailed specification*, which gives all dimensions for structures and provides a shopping list of equipment to be procured, installed. and commissioned by the contractor usually in the form of *bills of materials*.

Nominated Suppliers: Sometimes a purchaser will wish to nominate a particular supplier for certain items of equipment (e.g., pumps) in order to standardize spares and servicing. But "nominating" a supplier can mean contractual problems because a problem with that supplier's equipment would be the purchaser's responsibility not the contractor's. To avoid this, the purchaser will often issue a list of *preferred vendors*, which is not mandatory but alternatives may not be acceptable and the contractor who ignores preferred suppliers in favor of a cheaper alternative may find himself forced into providing the more expensive preferred equipment at his own cost.

General Specification: This section will refer to any appropriate standards such as BS (British Standards), CEN (European Standards), API (American Petroleum Institute), and ISO (International Organization for Standards) and provide copies of the purchaser's standard specifications for equipment, paint finishes, materials, and so on. It should be carefully examined for clauses such as "pump motors shall be 1450 rpm," which may increase the cost of equipment, often for no particularly good reason.

Program: Usually the contractor will be required to submit his proposed program for contract execution, but sometimes the purchaser will issue a program for the works and the contractor must establish if this is reasonable (the completion date is defined by cynics as the date on which liquidated damages commence—see Chapter 11).

Compliance: The contractor tendering against a specification is usually required to issue a statement along the lines of "our bid is in complete compliance with the specification" or to provide a list of all areas in which he has deviated from the specification.

Alternatives: The consultant who prepared the tender documents will have been paid a quite significant fee to produce a cost-effective design and specification. However, he may not be up to date with current technology or may decide not to investigate whether better options are available. He may allow the tenderer to offer an alternative design (provided that he submits a bid for the specified plant too).

Project Management: This section identifies how the project will be managed and often contains "hidden" costs associated with progress meetings, reports, quality assurance (QA) documentation, and health and safety issues. It might even stipulate what management software the contractor has to use (see Chapter 20).

Validity: The tender will normally be open for acceptance for a specified period of time or until a specified date, although the tenderer may revoke it at any time before it is accepted by the purchaser. Once the tender has been accepted by the purchaser the contract has been formed and the tenderer cannot revoke his offer even if he finds he has made a mistake.

Terms of Payment: The contract will set out when payments are to be made by the purchaser to the contractor, how much each payment will be, and what the contractor has to achieve in the way of milestones in order to be paid.

If the tender is submitted on behalf of a company, then it must be signed by an authorized signatory of the company—usually a senior manager or, if it's a major contract, a director.

THE ACCEPTANCE

In legal parlance, a tender is an *offer* and not an *invitation to treat* (no, this has nothing to do with buying drinks, *treat* here means to bargain over the price). This means that acceptance of the tender by the purchaser concludes the contract and is legally binding on the tenderer.

The tender may be conditional or unconditional but acceptance must be:

- Complete and unconditional and in the manner and terms of the offer— any suggested variation constitutes a counteroffer.
- Made within the validity period of the tender.
- Made while the tender has not been revoked.
- Accepted only by the addressee or his previously nominated agent.
- Communicated in writing—a letter, a fax, or an e-mail with a secure signature.

Note that acceptance of the tender by the purchaser concludes the contract and cannot be revoked. Because the drawing up of contracts often takes a considerable time, the purchaser may issue a *letter of intent* to allow the contractor to start work before the final contract is signed. This letter will normally be legally binding and will conclude the contract.

If there is some mistake or misrepresentation in the contract or if it was made under undue influence or duress, then it can be rescinded by a court. If this happens, the purchaser and contractor will be placed in the position they would have had if the contract had never started. This will involve the purchaser paying the contractor for the work he has completed. This is called *quantum meruit* (Latin for "as much as he deserves").

If there are no signed documents, a contract can still be made *by performance*. If the contractor starts the contract and the purchaser behaves in such a way as to imply that the contract exists, then the contract is deemed to have been made. If, for some reason, the contract then goes wrong, again the *quantum meruit* principle applies.

The rules that govern how the contract will be executed are set out in the conditions of contract (see Chapter 11).

SUMMARY

- **You hope there's someone in your company who knows all this.**
- **There had better be!**
- **It should be you.**

Chapter 11

Conditions of Contract

WHAT'S A CONDITION?

A *condition* is a term in a contract that effectively sets out a promise that the contractor or purchaser must fulfill. If the contractor fails to do what he promised, then the purchaser may refuse to pay; equally if the purchaser fails to provide something he promised to provide (e.g., site clearance), then the contractor may refuse to complete his obligations.

The conditions of contract are set down in order that each party to the contract understands what are his or her obligations and that these can be formally agreed upon and recorded. In particular they define:

- The *normal performance* required from each party (that's the work the contractor must do, the price the purchaser will pay, and how these may be varied within the framework of the contract without recourse to renegotiation)
- How the risks (design, commercial, delivery, etc.) will be apportioned between the parties
- The rules and procedures for conducting the contract and dealing with the problems and disputes that may arise
- The terms of payment

The conditions of contract are therefore fundamental to the execution of the contract.

MODEL FORMS

When engineering contracting first began, the purchaser appointed *the engineer* (see Chapter 23) to manage the contract, and he would write conditions specific for each contract. Later, as more contracts were placed and consulting engineering firms became established, it became convenient for each firm to have its own standard conditions. Naturally, these were invariably weighted

Engineering Money: Financial Fundamentals for Engineers, by Richard Hill and George Solt
Copyright © 2010 John Wiley & Sons, Inc.

in favor of the purchaser, so contractors began to write their own conditions, which were, of course, weighted in favor of the contractor.

As the contracting industry grew and projects became more complex, problems of misunderstanding, misinterpretation (that's deliberate misunderstanding), and, consequently, conflict between parties developed. As conditions of contract were expanded in the hope of covering all possible problems, they became lengthy and complicated. Worse still, one firm's conditions would be different from another's. In order to provide a consistent approach the UK's Institution of Civil Engineers (ICE) produced a "model form" of conditions of contract that could be used in all civil engineering contracts.

Other institutions felt that the ICE Conditions of Contract weren't appropriate to their trade. Thus, other standard conditions of contract have proliferated. Of the various different standard conditions commonly in use worldwide, many are those produced by the UK engineering institutions, which include:

- Institution of Civil Engineers (*ICE Conditions of Contract*, 7th edition, and *Design and Construct*, 2nd edition)
- Institution of Electrical and Mechanical Engineers and the Association of Consulting Engineers (IEE/IMechE)
- Institution of Chemical Engineers Conditions of Contract for Process Plant (IChemE)
- FIDIC (International Federation of Consulting Engineers) Conditions of Contract for Civil Engineering Construction
- Joint Contracts Tribunal (Building Contracts Conditions also called RIBA Conditions)

The situation is further complicated by the fact that there are different versions of the conditions for different types of contract. One of the reasons for this proliferation of conditions of contract is that the ICE Conditions of Contract aren't really applicable to mechanical and electrical plants. So, for example, there are:

Further complication: Tender documents often contain amendments and/or supplements to specific clauses in the standard conditions of contract. That's particularly common if the documents have been drawn up by consultants. They always think they are better qualified to write conditions than the institutions' committees.

- IChemE Red Book for lump-sum contracts
- IChemE Green Book for reimbursable contracts
- ICE Engineering Contract A: Conventional Contract with Activity Schedule
- ICE Engineering Contract B: Conventional Contract with Bill of Quantities
- ICE Engineering Contract C: Target Contract with Activity Schedule
- ICE Engineering Contract D: Target Contract with Bill of Quantities

The ICE conditions concentrated on tenders submitted against bills of quantities and remeasurement during construction, which gives quantity surveyors on both sides ample opportunity to submit claims and counterclaims. It's fairly obvious that working to conditions like this is confrontational.

Bills of quantities are appropriate in traditional civil engineering construction where, for example, large concrete structures are being built or long underground pipelines are being laid, and where the contractor is generally not responsible for the design. However, they don't work for process plant with pressure vessels, pumps, heat exchangers,

> Having spent most of my life in process plant contracting, I had had little contact with civil engineering. When I first came across quantity surveyors I asked a colleague what they actually did. "A quantity surveyor," he told me, "is the guy who goes round after the battle and bayonets the wounded."

and the like. What counts then is the performance of the process plant in meeting the required product quantity and quality, which is very much related to the process engineering skills of the contractor. The IChemE conditions attempt to address the problem by concentrating on process performance and by attempting to be conciliatory rather than confrontational.

Many countries have different versions of such standard conditions. The ones listed here apply to the UK, but they are used in many other parts of the world. To compile a worldwide list here is, of course, quite impracticable.

> "I have never seen an item covering a cubic yard of experience!" remarked a contractor complaining about this subject at a conference in 1971. Things have not changed significantly in the ensuing years.

SUBCONTRACTS

Few, if any, contractors have all the skills in-house needed to execute a modern construction contract, which requires civil, mechanical, electrical, and possibly process engineering skills, not to mention possible specialist skills such as orbital welding of stainless steel or the supply of specialized items such as pharmaceutical stills. A *main contractor* needs specialist subcontractors to provide these. In fact many of the large main contractors employ no direct labor themselves but act as *management contractors*, and rely on subcontractors for all design and site labor.

This means that the conditions of contract between the main contractor and subcontractor are of great importance. A subcontractor may be solely responsible for a delay to the whole contract, but it is the main contractor who is responsible for the execution of the contract and therefore liable for the full liquidated damages—the subcontractor can argue that the main contractor's mismanagement has resulted in the delay. So, in practice, the main contractor tries to pass responsibility and risk down to subcontractors, Main contractors generally try to impose on the sub-contractor the same conditions of contract that apply to his contract with the purchaser, particularly in respect of *damages* (see below). This *back-to-back* arrangement is like a bookmaker laying off bets.

To avoid this sort of dispute and to make each subcontractor carry an appropriate part of the risk, most main contractors have their own conditions of contract applicable to subcontracts.

Passing risk to subcontractors means that a painting subcontractor with a contract valued at, say $50,000 could be responsible for delaying a $100 million oil refinery and be faced with paying liquidated damages of up to, perhaps, 5% of the main contract price or $5 million. This is clearly unfair and quite ridiculous. On the other hand, if the painting contractor is solely responsible for the delay, the main contractor still has to pay the full liquidated damages.

FORCE MAJEURE

Force majeure (French for superior force) is a common clause in contracts that essentially frees both the purchaser and contractor from liability or obligation if an extraordinary event occurs or if circumstances beyond the control of either party prevents one or both of them from fulfilling their obligations under the contract. Such events include war, strike, riot, crime, or an event described by the legal term *act of God*, for example, earthquake, flooding, or volcanic eruption.

However, *force majeure* is not intended to excuse negligence or sabotage by one of the parties or where nonperformance is caused by the usual and natural consequences of external forces such as normal weather conditions.

DAMAGES

The conditions of contract will usually set out the guarantees that the contractor is required to meet. If a contractor fails to complete a contract on time or, in the case of a process plant, if the plant fails to meet the required performance, then the purchaser may suffer a loss and the contractor would be liable to pay damages in compensation for that loss.

The damages may be *at large*, that is, they will be a direct reflection of the amount of money that the purchaser has lost. Suppose a contractor builds a factory and, because the foundations were faulty, a wall collapses and has to be rebuilt. This would constitute a warranty claim and the contractor would be liable for the cost of rebuilding. Now suppose that the wall collapsed onto an expensive piece of machinery. The purchaser has now suffered a *consequential loss* as result of the contractor's mistake. The contractor could be held liable for the cost of repairing the machine unless the contract made it clear that he would not be held so liable.

As another example, suppose that a factory was due to be completed in time to produce toys for the Christmas market and, through the contractor's fault, the factory was not completed until January. The purchaser can justifiably claim consequential loss. He may be able to estimate what his sales would have been and so quantify how much he has lost because of the contractor's failure. He could sue the contractor for this sum but would have to justify the sum claimed.

In process plant contracting the situation is even more difficult. Suppose that a contractor guarantees that the power consumption for a particular item of plant will be 10 MW, but, when the contract is completed, the *performance test* shows that it is actually consuming 11 MW. At, say 5¢/kWh it will cost the purchaser around $200,000 extra per year to run the plant. This is a clearly identified loss for which he might justifiably claim damages from the contractor. If, as a result of the additional power costs, the product the plant is producing is too expensive to sell or the profit margin is reduced, then, again, the purchaser has suffered consequential loss and could claim damages from the contractor: They could run into millions of dollars.

> Here's a real-life example: In 2005 a UK supermarket began a project to build a new store above a railway cutting that involved a tunnel of precast concrete segments over the rail line. The tunnel partially collapsed, blocking the rail line, and services were disrupted for several months.
>
> The initial loss was the cost of clearing the railway line and reinstating the collapsed tunnel section. However, there was a series of consequential losses including:
>
> - The opening date for the store was delayed resulting in loss of projected income.
> - The railway company lost income because the track was unavailable to run trains.
> - The railway company had to meet the cost of providing buses for their customers as an alternative to trains.

Assessing, justifying, and claiming such damages at large is a lengthy and expensive legal process, and the concept of *liquidated damages* was introduced to avoid it. It is a sum that is agreed to in advance by the parties as being a reasonable assessment of what the purchaser's loss would be. The sum might be based on a percentage of the contract price per week for delays, or on a formula for calculating financial penalties for shortfalls in guaranteed figures such as product quality or chemical or power consumption. Note that if the sum agreed is excessive, then, in law, it is a *penalty*—that is, a threat—and that is unenforceable in most legal systems.

A contract may contain the term *time is of the essence*. This is a legal term that specifically means that the contract must be completed by the time specified, otherwise the purchaser is entitled to reject the entire works and recover all the costs incurred, even if the contractor is only a day late in completion. This could easily bankrupt the contractor.

As well as guarantees for performance, contract conditions will usually identify a *defects liability period* or warranty, that is, a period of time—typically 1–3 years—during which the contractor is liable to replace or repair any part of the works that becomes defective. This may include items of machinery, such as pumps, which have been

> We bought some pumps too early in a contract that was then delayed by 6 months. By the time we started commissioning the plant the 12-month warranty on the pumps had expired. Two of them failed during commissioning, but, because the warranty had expired, we had no claim against the supplier and had to pay for new pumps.

supplied by the contractor and which carry a normal manufacturer's warranty. If the period of this warranty does not cover the whole of the defects liability period, then the contractor may be at risk.

The contractor's liability will also cover *latent defects*, that is, faults in materials, workmanship, or design that are not immediately apparent but that become apparent during the defects liability period. Suppose a pump that draws water from a concrete reservoir fails and is repaired under warranty. Suppose it fails again and it becomes evident that the cause is a badly designed suction arrangement in a concrete tank. The design-and-build contractor will be liable not only for replacing the pump (probably a relatively small cost) but, potentially, for rebuilding the reservoir.

With construction projects becoming larger, growing environmental protection legislation, and other such factors, the risk of consequential losses becomes ever greater. It is, therefore, increasingly common for contractors to have to provide financial guarantees to cover them. This means that only large highly capitalized companies will survive as main contractors and explains why there is more amalgamation in the construction industry leading to fewer larger players.

DISPUTES AND SETTLEMENT

The climate of contracting has become much more confrontational for two reasons. First, contracting has become so competitive that some contractors are knowingly tendering prices below breakeven, in the hope of making their profit by negotiating "extras" during the contract. Second, the world is following the lead from the United States in becoming more litigious both in terms of disputes between main contractor and purchaser and between main contractor and

> I was an expert witness in an action where a small electrical subcontractor from a country town claimed over $2 million worth of extras against a very big main contractor. He was encouraged by his solicitor (who could see large fees for himself) in spite of the fact that the main contractor's solicitors were a large London firm. The electrical subcontractor's claims were grossly overestimated and the claim was settled out of court for $400,000. The legal costs came to more than this and the electrical subcontractor went bust.

subcontractors. But remember that litigation is expensive, especially if you lose, and lawyers for big companies can slow the process down so much that their smaller opponents have to give up and *settle out of court* before a conclusion is reached.

In general, it is worth remembering that, conditions of contract and specifications notwithstanding, in most legal systems the contractor must supply goods that are *fit for purpose,* and conditions of contract that are deemed unfair by a court are not enforceable in law.

When disputes occur, the conditions of contract should set out the procedures to be followed. Litigation is expensive for both sides in a dispute and frequently takes months or even years. This is particularly hard on small subcontractors, whose payments may be delayed until main contract claims have been settled, and who can be ruined as a result. Most conditions of contract, therefore, try to resolve disputes without recourse to the courts.

In normal day-to-day operation, the project manager, who is usually nominated in the contract, or his delegated agent is the arbiter of the conditions of contract. (In some conditions of contract the project manager is called "the engineer," but this is generally recognized as adversarial.) Work to be carried out under the contract is often required to be *to the satisfaction of the project manager.* If there is a dispute between the contractor and the purchaser, then the project manager will normally rule until the dispute can be settled by a third party.

Under European legislation any dispute under a contract that includes construction operations can be referred to an *adjudicator* who will ascertain the facts of the dispute and give a ruling within 28 days. That ruling will then bind the parties unless reversed by arbitration or litigation. Under certain conditions of contract (e.g., IChemE) the role of the adjudicator is carried out by a mutually agreed *expert*. These schemes aim at a rapid low-cost settlement.

If adjudication fails, the next step is *arbitration*, where an independent *competent person* (nominated under the contract) is called upon to resolve the dispute. Arbitration is a legal procedure involving lawyers and the discovery of evidence, and therefore costs more time and money than adjudication. However, if arbitration fails, the only alternative is litigation, which is more expensive still, and beyond the means of most small contractors.

> *"In England, Justice is open to all—like the Ritz Hotel."*
>
> —*Lord Justice Mathew (1830–1908)*

HOW TO AVOID DISPUTES

Long experience in this dismal subject shows that most of the disputes arising out of contracts are the result of misunderstanding or misinterpretation of the specification. So it's absolutely vital that engineers write clear and unequivocal specifications.

SUMMARY

- More legal stuff, but of crucial importance to anyone involved in project engineering, whether as purchaser of contractor.
- A good relationship between the project manager and the contract manager is the best way to stay out of trouble.
- When major problems seem impossible to resolve, it is usually because the conditions of contract are poorly drafted or unclear.
- Ultimately contract conditions are a matter for lawyers, but engineers need to understand them so as to be able to talk to the lawyers.

Chapter 12

How Things Can Go Wrong—2

This is the story of a large gas-fired power station that was built to supply power to a publicly owned industrial site. A joint venture company (which I shall call *the Company*) had been formed for the project whose shareholders were the industrial site, the contractor who was to build it, and an electricity company that would run it—a DBFO (see Chapter 17) scheme. The project went horribly wrong and I got the job of expert witness in the huge lawsuit that resulted.

Expert witnessing always starts with the arrival of a small truck load of files and papers that must be read to get the background, and this is what I found:

Construction of the project had been badly mismanaged, and the power station was more than 6 months late coming on line. That not only meant extra costs but also several million dollars worth of lost revenue. The Company decided to make the bank pay out under the contractor's performance guarantee. The only way in which the contractor could recover the money was to sue the Company, which is the normal pattern in these sad events.

I wish I could have been a fly on the contractor's boardroom ceiling at the board meeting when they decided to sue the joint venture company in which they were a partner. It must have been a lively affair because, of course, some of the directors on the contractor's board were also board members of the joint venture.

Mistake 1 was that the Company had handed the contractor responsibility for design and construction without either going out to competition or taking on a consultant to provide proper control. Such a buddy-buddy relationship lacks the rigor necessary for good engineering. I found that no proper plan for the execution of the whole project had ever been drawn up, and, of course, there was no critical path analysis. The worst thing was that the contractor built the plant as if he had won the job by competitive tendering, when a

Engineering Money: Financial Fundamentals for Engineers, by Richard Hill and George Solt
Copyright © 2010 John Wiley & Sons, Inc.

contractor's objective is to meet the contract's specification at the lowest cost to himself. The plant he was building would actually belong to the Company in which he had a big shareholding, so the contractor should have cared more about its operating problems than his immediate construction costs.

Modern boilers and steam turbines need very pure feedwater, and the production of that calls for specialized chemical knowledge. The contractor had put his Mr. Silly, a switchgear engineer, in charge of the subcontracts for the boiler and turbine sets, which included the water treatment plant. I checked on Mr. Silly's background and found he had never studied chemistry: he was out of his depth, but wouldn't admit to it.

At the preliminary enquiry stage Mr. Silly presumably just consulted the textbooks and issued an enquiry specification for the plant based on a conventional and simple process. The proposals engineer of one of the potential subcontractors for this plant was an old friend of mine. He told me that the impurities in the water supply to the site made the water difficult to treat. He had warned Mr. Silly at a pre-tender meeting that his enquiry specification was not up to it and was told to "**** off and quote in accordance with the specification!" That's what my friend did, and so did the other companies that competed for the contract, though they knew that if that design were built, it would fail. "It happens all the time" he told me sadly. "We often have to quote to a specification even when we know it won't work in practice."

So mistake 2 was failing to admit a lack of specialist knowledge and ignoring professional advice. I suppose Mr. Silly was suspicious because my friend was trying to sell him something, but then he was an expert in this specialized field, and Mr. Silly wasn't.

At this stage, the water purification plant subcontractors paid only limited attention to Mr. Silly's enquiry because Mr. Silly's company was itself only tendering for the project and didn't actually have an order to place. Mr. Silly chose one of the offers (not from my friend's company) to go into their tender for the whole project, and the contract between the contractor and the Company was duly signed on that basis.

Now Mr. Silly reissued the same enquiry document as before. This time he had a real order to place, so the subcontractors treated it much more seriously. As it happened another potential supplier offered a plant design that was considerably smaller (and cheaper) than the one that had been included in the main contract papers. Mr. Silly ordered it. This was mistake 3.

I was puzzled by that—the contractor's offer to the Company clearly included the bigger plant, and the contract document even gave its dimensions, so this seemed to me an obvious breach of contract. The Company had engaged a large consultant company to supervise the project, and they should have spotted the discrepancy. But it turned out that the bank, on whose loan the project depended, had insisted on the consultant being engaged (to protect its investment, it was thought) and the Company had actually made this a rather nominal arrangement. The fee actually paid to the consultant was so low that it only paid for minimal hours to be spent on the job. Minor details of this kind had simply escaped notice. Now that the contractor was suing the Company for his money, the Company was hardly going to point out its own shortcomings.

Mistake 2 meant that the plant design was flawed and it would not do the job: Mistake 3 now meant it was also going to be too small. That became apparent during the construction period and getting bigger units delayed startup for 6 months.

Adding an additional treatment stage to overcome mistake 2 did not take very long. It just cost more money.

No one attached any blame to the subcontractor who had built the undersized and inadequate plant. He had quoted to the specification and Mr. Silly had accepted his offer. The water treatment plant subcontract specified in detail what was to be supplied, and it was supplied. If it didn't satisfy the need, that wasn't the subcontractor's fault.

One morning we were in the lawyers' offices working away on the papers for this lawsuit when we were told to stop work: The parties had settled out of court. This is the usual and sensible way in these cases—actual court appearances are hideously expensive and therefore to be avoided. The details of settlements remain confidential, but there is a telling clue to this one: The contractor went out of business a few days later.

SUMMARY

- **DBFO only yields better engineering if the plant is built with a view to its future performance, rather than the lowest cost to the contractor.**

- **Main contractors, and even some general consultants, just don't have the know-how to cover some specialized aspects. Specialized subcontractors or specialized consultants do.**

- **If one design to meet a specification is similar but much bigger than another, one of them must be wrong. As nobody overdesigns in competitive tendering, it is probably the smaller one.**

- **Consultants have an important function—if they're given a chance to perform it.**

- **If you've been given the job of writing a specification and you don't have the necessary technical expertise, it's best to seek help.**

Chapter 13

Cost Centers

WHAT'S A COST CENTER?

A cost center is an accounting tool that allows us to take staff costs, which would otherwise be regarded as part of the company's overhead, and allocate them to contracts and thus convert them into direct costs.

When a contractor makes up a price for an offer, the biggest item after the estimated BOC (bought-out costs including material and services) is almost always the contribution, most of which is needed to cover the company's overhead. Just to remind you: Overhead is the costs that can't be attributed to a particular contract. If we could attribute more of the costs to contracts, therefore, the direct costs would be higher and the overhead lower, so that the percentage of contribution that has to be added to calculate the selling price would be smaller. That, in turn, would yield a more accurate costing and, therefore, a more accurate estimate for the price at which the contract should be offered. It is important to remember that this is an accountants' device—on its own it changes nothing.

HOW DOES IT WORK?

Many of a company's full-time employees do work that could indeed be attributed to contracts, which is what cost centers are created for.

To take a part of a company and create a cost center from it, you need to be able to do two things. First, you can only do it if it is a part of the company whose work and costs can be separated out from the rest of the company. Second, you must be able to measure the work that it

> All this modern management stuff didn't exist in the 1960s when I worked for a process plant contracting company. The accounts system had no cost centers, so none of the work done within the company could be booked to contracts. As a result, the company's standard price makeup formula was BOC +50%—wonderfully simple and wonderfully inaccurate!

does for the company. It is rather as if you were creating a separate organization within the company.

For example, given the irregular timing of contracts, my company could never guarantee that our highly skilled erection and commissioning team (see Chapter 7) would always have work throughout the year. When they didn't, we would hire them out to other companies, which made it important for us to know what their work was actually costing us. So we set up a cost center.

> If you create a cost center for some department, it will show more clearly what service the company receives from the department, and how much that costs. Once the company's management has that information, they might well be persuaded that others could provide the same service at a lower cost. Outsourcing starts here.

MEASURING THE COST

So the erection cost center was treated as if it were an independent company, but first we had to find out what the whole thing cost. The biggest single cost was, of course, the salary cost of all the staff, including those who weren't in the front line, such as the department manager and clerk. "Salary cost" includes all associated costs such as taxes, insurance, and pension. The department's costs, such as transport, telephones, and IT—all quite high for this type of work—were added. A more sophisticated scheme might also have added rent for their office space, electricity, and so on. By adding all these together, we got the total sum it cost the company to run the erection department for the year.

MEASURING THE OUTPUT

The department's output was measured in terms of person-hours. We know how many operatives (the people who actually do the work) we have, but how many hours of attributable (money-earning) output can we expect them to perform?

Let's assume that the staff works 40 hours a week. If we allow 4 weeks for vacations and 2 weeks for national holidays, that makes 46 weeks, or 1840 working hours per year. The real number, however, is much lower because there are always a lot of hours that can't be attributed to a specific item, for example, time spent on remedial work for an old contract whose books are closed. As a practical figure 1500 hours/year is more realistic.

We can, therefore, calculate that the cost per person-hour is

$$\frac{\text{Total department cost}}{\text{Number of operatives}} \times 1500$$

When preparing offers, our estimating department had to estimate the person-hours the erection of the project would take. The cost center gave us a cost per person-hour, so we could now calculate an estimated erection price to go into the price makeup as a separate estimated item. Without the cost center, it would have been an unknown amount lumped together with all kinds of other costs in the contribution.

Project management, drafting, and commissioning can all be treated just the same, and the output of those cost centers, too, would be in person-hours. Engineering is a bit more complicated because the department is usually bigger and multidisciplinary, and there is considerable variation in salaries and overhead costs across those disciplines. For our first few years we considered all design work as a single cost center, but it became clear how big the differences were, and we treated each discipline separately with its own cost per person-hour.

Obviously, the system depends on timesheets for anyone whose person-hours are being attributed to specific items. There is a range of levels of sophistication—our Dutch partners went overboard and installed a system in which everyone, including the CEO, filled in timesheets. I thought this unreasonable since the numbers for calculating costs are based on estimates, and there's no point in aiming at great accuracy. It means more form-filling and costs more than it's worth.

The output of a cost center isn't necessarily measured in terms of person-hours. Purchasing, for example, is more readily evaluated in terms of the number of orders to be raised on the contract. The total cost of the purchasing department is divided by the total number of orders placed to give a cost per order.

The company I quoted above for having to price its contracts on the basis of BOC plus 50% went down the drain and several of its former staff (including me) founded a new company to undertake the same kind of contracts. We installed a management accounts system that used cost centers and that made our price makeup formula BOC plus cost center costs plus 25%. In other

Cost centers also provide a way of monitoring efficiency. When the person-hour cost of our drawing office increased above that of our local drafting agency, we had to consider whether it was actually worth keeping our own computer-aided design (CAD) operators or simply outsource the service. (We decided to keep our own staff but managed to reduce the department's costs.)

words, where the old company could estimate only about half the contract price with any hope of accuracy, the installation of cost centers brought that up to three quarters. All estimates are, of course, more or less inaccurate, but they are better than nothing. It meant we could gauge the selling price more accurately, which is crucial in any competitive industry. The downside of this is that it means more administration for everyone, and it can get quite complicated.

THE ADMINISTRATION

A contracting company will have engineers who are permanent staff employees, but it will often take on agency employees to provide additional resource for a specific contract. They will work in parallel with the permanent staff who may be seconded full-time to a single contract, or may be working on a number of contracts or proposals and sales (see Chapter 14).

The agency engineers are a direct cost to the contract, which is straightforward. Staff engineers are, in reality, part of the company's overhead, but their time (whether full time or part time) on a contract is "bought" by the contract manager from the engineering department cost center and the cost debited to the contract.

Keeping track of this requires a management accounts system in which most staff complete weekly time sheets in order to identify the number of person-hours they have spent on particular contracts. So, for example, a structural engineer who is a staff employee may allocate 8 hours to one contract, 16 to a second, 4 to a third, and 4 to proposal preparation. If he works a 35-hour week this leaves 3 hours unaccounted for during which time he probably reads some journals to keep up to date with technology or provided advice on the telephone to various other people. He will probably allocate this time arbitrarily to one of the contracts in order to appear fully employed, but he shouldn't.

> If you are a staff engineer working on a number of projects at the same time, it is often difficult to keep track of time spent on each (or none). The time sheet should note the detail of how the day was spent, but the result is often a work of fiction (science fiction?). The form on which the information is gathered usually has various codes for contracts and other activities. I once worked for a research company that had a cost code for "reading and thinking unrelated to specific projects." We had one engineer who booked the majority of his time to it.

Time sheets are universal to all kinds of industry, not just project engineering, but they are particularly useful where the work covers various objectives and activities.

When you install such a scheme, you have to decide how sophisticated to make it. I mentioned schemes in which the cost center is charged with rent, heating, and electricity, and our Dutch partners where even the CEO had to fill in a time sheet. We thought that to go as far as that would create paperwork whose nuisance outweighed the gain. Given that the results could never be very accurate, it seemed silly to impose extra bureaucracy on everyone.

A MONITORING SYSTEM

For cost centers to improve the accuracy of cost estimating, the company must be able to estimate the person-hours each cost center is likely to spend on a particular contract. To do that needs experience and a library of records from previous jobs.

That in turn leads to another benefit from cost centers. They have another use, which is perhaps even more important than greater accuracy in setting the selling price.

When estimating the person-hours for a job, one can also predict when they are likely to be carried out. Suppose, for example, that a job is expected to take a year to complete. The time estimate consists of building up a total from the various activities as they take place during the year. That might suggest that 500 hours are needed in the first month, 300 in the second, and 100 hours in each of the next 7 months. These estimates go to management accounts who record the actual person-hours booked each month, side by side with the original plan.

If the hours used vary significantly from the plan, it shows that something is going wrong (it could, of course, be the estimate itself). If the figure is too high, there seems to be trouble somewhere: If it is too low, it means there has been some unexpected delay. Either way, it signals that something probably ought to be done. That's why it is important for the management accounts people to be prepared quickly—if they come out 2 weeks late, it is probably too late to do much about any problems that exist.

Thus, the cost center not only improves the company's ability to price its offers, but it leads to a system that helps monitor the work as it proceeds. The accounts also provide historical data of person-hours spent on contracts against those estimated, which builds up the library and improves estimating person-hours for future contracts.

Management accounts will also identify the actual cost of departments and how much they are *recovering* in terms of person-hours multiplied by the person-hour rate or whatever other basis is used. Our practice was that if there were a significant discrepancy by midyear, then the rate was probably wrong and was adjusted.

SUMMARY

- **Cost centers are a way of converting staff costs from overhead to direct costs.**
- **They provide a means of calculating a cost per person-hour (or some other criterion) for each department.**
- **They provide a way of achieving better estimates of contract costs for pricing.**
- **They also provide the basis for monitoring progress, managing cash flow, and estimating future contracts.**

Chapter 14

Pricing Contracts

PRICE OF THE CONTRACT

When I first started work in the proposals department of a contracting company an older, wiser, and very much more cynical engineer told me that a contract price was "a blind guess carried out to eight significant figures." I thought he was joking but, looking back over nearly 40 years, I can see that there was more than a grain of truth in it.

The fact of the matter is that cost, price, and value all mean rather different things, although they are all expressed in the same number, which, in our case, is the contract price. Contracting is a competitive business, and, all other things being equal, purchasers generally place contracts with the lowest bidder. So, just as in the supermarket, the lowest price gets the order. So too high a price means that the company doesn't win any contracts and fails because it cannot cover its overhead, and too low a price means that the contribution is too low to cover the overhead and the company will fail. But, equally, the lowest price doesn't always mean that the purchaser is getting the best value.

COMPETITION

The story of how we founded our company (see Chapter 6) shows how relatively easy it is to start a contracting company. We were not the only ones to discover this, so new and hungry competitors were always springing up.

With such fierce competition, profitability (i.e., profit as a percentage of turnover) will always stay low, but, provided that the potential return on capital is high, this is not a business problem. On the other hand, low profitability means that the margin for error is small, and contracts can easily swing into loss. In many real-life cases it only takes one or two bad contracts to break a company—even a big one.

No two contracts are the same. Even if the technical content of a contract is the same as that of a previous one, there will be different site conditions, labor conditions, costs may have shifted in the interval, the time of year and

Engineering Money: Financial Fundamentals for Engineers, by Richard Hill and George Solt
Copyright © 2010 John Wiley & Sons, Inc.

the weather play a part, and so on. Compare this to manufacturing, which takes place in a more controlled environment, making it easier to give an accurate estimate of the cost of producing the product.

MAKING UP THE PRICE

When a contractor receives an enquiry from a purchaser, there is a limited, and usually short, period in which he has to submit his price. During that time (typically 6–12 weeks) the contractor's proposal manager has to get a preliminary design prepared, estimate the materials and subcontract costs (usually called the *bought out costs*) and the direct costs from his own cost centers—engineering, procurement, project management, and so on. To this he will add any other direct costs and finally the contribution calculated in the year's business plan (see Chapter 7). This will all be set out in a document called a price makeup sheet or something similar, like the one that we looked at before (see Chapter 9).

Now is the time to have another look at it and see what the numbers mean. The price makeup sheet is a very important document because it not only shows how the tender price was arrived at but, if the company wins the contract, it will be the basis of the contract manager's budget and cash flow forccasts.

The form I have set out in Table 14.1 is a simplified version of a real price makeup sheet. Of course, every company has its own way of setting things out, but this will serve to illustrate the important points. It represents a typical mechanical and electrical contract of the type that my company used to carry out, and it will be a good dcal more complicated for a civil/mechanical/electrical turnkey contract. The price makeup sheet is a summary and will be supported by detailed estimating sheets and quotations from material suppliers and subcontractors.

The following notes expand on Table 14.1:

1. The materials cost is the sum of all the equipment and materials that will be purchased for the contracts—concrete, shuttering, steelwork, pumps, pipes, specialist machinery, and so on.

2. Subcontracts are packages of work that will be carried out by other companies on site as opposed to direct labor employed by the contractor. Typically, this might include piling, scaffolding, equipment installation, electrical cabling, painting, and similar activities. The sum of materials and subcontracts is the bought out cost (BOC) and is usually the biggest single cost item.

3. Engineering is estimated on the basis of experience, that is, the proposal manager will talk to the various discipline engineers, explain what's involved in the contract, and ask them to estimate how many person-hours of design they will need.

Table 14.1   Contract Price Makeup Sheet

Item	Quantity	Rate	Amount	Total	
Materials-mechanical			$950,000		
Materials-electrical			$500,000		
Materials-control & instrumentation			$200,000		
Materials deliveries			$20,000		
Sub Total Materials				**$1,670,000**	Note 1
Mechanical installation sub contract			$100,000		
Electrical installation subcontract			$60,000		
Sub Total Installation				**$160,000**	Note 2
Process engineer	300 hrs @	$120.00 /h	$36,000		
Mechanical engineer	650 hrs @	$100.00 /h	$65,000		
Electrical engineer	450 hrs @	$100.00 /h	$45,000		
Controls engineer	400 hrs @	$120.00 /h	$48,000		
Drafting	500 hrs @	$80.00 /h	$40,000		
Sub Total Engineering				**$234,000**	Note 3
Commissioning	1200 hrs @	$100.00 /h	$120,000		
Sub Total Commissioning				**$120,000**	Note 4
Conract Manager	2000 hrs @	$120.00 /h	$240,000		
Procurement	500 hrs @	$80.00 /h	$40,000		
QA	300 hrs @	$100.00 /h	$30,000		
Safety	400 hrs @	$100.00 /h	$40,000		
Sub Total Management				**$350,000**	Note 5
Site offices	50 weeks @	$800.00 /week	$40,000		
Craneage	200 days @	$800.00 /day	$160,000		
Support staff	8000 hrs @	$60.00 /h	$480,000		
Telephone			$20,000		
Stationery/printing			$20,000		
Transport			$20,000		
Sub Total Site facilities				**$740,000**	Note 6
Interest	2.0% of Contract Price		$110,050		Note 7
Agents' fees	1.0% of Contract Price		$55,025		Note 8
Fixed price	1.0% of Contract Price		$55,025		Note 9
Licence fees	0.5% of Contract Price		$27,513		Note 10
Insurances	1.0% of Contract Price		$55,025		Note 11
Contingency	5.0% of Contract Price		$275,126		Note 12
Sub Total other costs				**$577,765**	
Total Direct Costs				**$3,851,765**	Note 13
Contribution	**30.0% of Contract Price**			**$1,650,756**	Note 14
CONTRACT PRICE				**$5,502,521**	Note 15

4. Commissioning costs are estimated in much the same way but should include any consumables and special testing equipment that the contractor has to supply.

5. Contract management costs are estimated in much the same way.

6. Site costs will vary from contract to contract. Here good design can save a lot of money. Prefabrication off-site might mean the difference between having an expensive tower crane on-site for months and having a mobile crane on-site for days (see Chapter 20). Small things that are often overlooked that can increase costs—site safety clothing, computers for the site office, and photocopying vast amounts of paperwork for QA records and cost analysis. Large contractors will own earth moving equipment, vans, site cabins, computers, and so on and simply rent them to the contract manager for the contract period. Smaller companies will have to buy or hire this sort of stuff.

7. At the tender stage, the proposal manager will prepare a cash flow forecast and an S curve (see Chapter 21) for the contract so as to be able to estimate how much money will have to be borrowed and when and how much interest will have to be paid.

8. In overseas contracting it is quite usual to have an agent (an individual or a company) who knows the local situation and has good contacts within the purchaser's company. Thus, his help may be invaluable in winning the contract.

> Bribery is clearly illegal but, in some countries, it was a way of life and business could not be done without it. Some of the contracts we tendered in Africa back in the 1970s contained high levels of "agents fees". Nobody knew what the agent did for that money.

9. Most contracts are *fixed-price* contracts, which means that the contract price will not change during the contract period. However, since the BOC is estimated on today's price, it may increase during the contract period. Even in times of low inflation there is likely to be a small increase in material costs and quite probably engineering costs, during an 18-month contract, so either the suppliers have to quote on a fixed price basis or a percentage has to be added to allow for likely increases.

> Don't confuse fixed price with "price fixing". What people generally understand by price fixing is an illegal practice, when contractors get together secretly and rig their prices high: The one who gets the contract then pays off the others. That is legal only if it's done openly and with the client's consent, which is rare.

10. In some specialist sectors, contractors might be using technology that they have obtained under a license agreement, for which they have to pay a license fee or royalty to the owner of the technology.

11. Contractors normally carry a *contractor's all-risk* insurance that covers them for damage or injury to staff, subcontractors and third parties as well as for mistakes in the design and consequential losses. Sometimes they need to take up special insurance. For example, overseas contracts may involve insuring equipment in transit by sea. The cost of these special insurances can be quite significant.

12. Adding all these items together gives the total direct costs of the contract.

13. The contribution is, of course, a key number. It may be the straightforward number calculated in the year's business plan or it may be higher if the contractor is confident of winning the contract or lower if he wants to be sure of getting it.

14. The price may include a contingency to reflect a particular type of risk or liquidated damages (see Chapter 10) in the contract or to provide a *negotiating margin* for the salesman to give as a discount without affecting the overall return on the contract.

15. And there it is: the contract price—well almost.

Note that profit isn't mentioned anywhere. A single contract makes no profit for the company, so it's senseless to speak of profit at this stage.

FINALIZING THE PRICE

The price makeup sheet is not the final step. There is the *price settlement meeting*.

All the things I've talked about so far are costs, and they are quantifiable. Now we have to look at a set of factors that influence the price but are not quantifiable. These are discussed at the price settlement meeting (other companies will have other names for it) when the people directly involved in the tender—the sales engineer, proposals engineer, contract manager, engineering manager, and at least one director—will discuss the factors and make a final decision on the price to be submitted. So what are these unquantifiable factors?

The first is competition. The sales department should act like an army's intelligence corps and try to find out as much as possible about the background to the project. In standard commodities like automobiles or cans of beans, you look at

In the 1990s the UK economy was in recession, and there was little construction work going on except in water works, which had to upgrade to meet new legislation. This sector became very competitive and one contractor in particular won most of the work by offering very low prices. Soon other contractors refused to compete with him, and the water companies soon found that they were unable to get three contractors to bid for contracts. This forced them to place contracts with other contractors who weren't necessarily the lowest priced.

the competitors' price and work to that. In the motoring magazines you can read that such and such an automobile is aimed at the Honda Accord market—that is, the lower priced, small-sized, five-door family car. Other manufacturers know the cost of an Accord and design and price their models to compete with that. In competitive tendering this is (theoretically) not the case, though there are devious means of finding out what the competitors' prices are.

Does the purchaser have a budget limitation? If so, there is no point in tendering a price that is higher. Does the purchaser have a preference for a particular technology that is available to only one of the bidders? Which other contractors are tendering and do they have a particular reason for submitting a low price—are they short of work, for example? Some contractors have a reputation for bidding low prices.

Second, there may be a particular reason why a contractor might want to win a particular contract. Perhaps it's a "prestige" project that will enhance his reputation, in which case he might want to offer a particularly attractive price.

Third, it's important to be aware of any particularly onerous conditions of the contract—punitive liquidated damages, for example, or a particularly high risk such as unknown ground conditions, contaminated land, or a site of special scientific interest—that will require special contingencies to be added. Contracting is a risky business, but risk can be minimized provided that money is available.

Weighing all these factors will usually lead to an adjustment to the calculated price—typically the sales department will be arguing to reduce the price whilst the operations department, which has to execute the contract, will argue for a higher price. And there it is: A blind guess carried out to eight significant figures.

Back in the 1980s there was a sudden growth in the semiconductor industry in Scotland. Each new factory needed a plant to produce ultrapure water. The technology was poorly understood and shrouded in mystery. It was important for contractors to have demonstrable experience in the field before they could win a contract. The five or six specialist contractors each "bought" a contract by tendering a price with little or no contribution in the hope that, having built one plant, the experience gained would stand them in good stead for the next contract, which they would price normally. By the time each contractor had done this, there were no more factories to be built, and none of them made any money out of what should have been a premium price market.

COST OF TENDERING

Preparing tenders is a costly process. Although the tender stage design is not fully detailed, it will take a full team of discipline engineers—process, mechanical, electrical, and civil—as well as planners, estimators, architects, and draftsmen. A typical $5.5 million tender might cost $50,000 to prepare—that's 1% of the contract price, and that cost is part of the contractor's overhead. Put another way, the cost of tender preparation is nearly 3% of the anticipated contribution on the contract. We used to expect to win one tender in four, which I think is fairly typical, and we got no reward for the work we put into the other three unsuccessful tenders.

> In Holland it's not uncommon for contractors to club together and agree on a sum which covers the cost of all of them tendering. Each offer includes this sum, which is then shared out between them out of the winner's contract price. If this is done openly, it is quite legal, but I have never heard of it being done anywhere else. The Dutch are a very sensible nation.

SUMMARY

- **Many items go into working out a contract price.**
- **We try to minimize uncertainty by using cost centers and so on, but much remains a best guess.**
- **With small profit margins, small errors make a big difference to profitability.**

Chapter 15

Competitive Tendering

TENDERING

In competitive tendering, a purchaser issues an enquiry for a project to a number of contractors. They make their offers, of which he selects the one he thinks is best and awards that contractor the order. We have already looked at the steps by which a project proceeds. In competitive tendering the key steps are:

- The contractors make their offers: their objective is to do that at a lower price than their competitors'.
- The purchaser's objective now is to get a satisfactory product at the lowest price: he awards the contract to the contractor whose offer appears to him to be the best.
- The contractor does the job: his objective is to meet his contractual commitments as cheaply as he can.

The clash between the purchaser's and contractor's objectives is clear. Naturally enough, the contractor wants to get the most money for the least expenditure and, just as naturally, the purchaser wants to get the best product, also for the least expenditure. Theoretically, the two aims can both be met, but competitive tendering is not always a good route for getting there: it does not give purchaser and contractor a common objective.

The purchaser (or quite often his consultant) prepares a document (an *invitation to tender* or *tender document*—the word *tender* simply means an offer) and sends it to prospective contractors. The tender document includes a specification for the work to be executed, together with conditions of contract and technical information. Each contractor prepares a document (a *proposal*) that sets out how he will achieve the requirements of the specification and includes the price he will charge the purchaser for carrying out the work. The purchaser reviews the tenders and selects that which represents the best value. With a detailed specification this is usually the lowest price, but issues of reliability and so on can arise.

Engineering Money: Financial Fundamentals for Engineers, by Richard Hill and George Solt
Copyright © 2010 John Wiley & Sons, Inc.

In many countries around the world the law obliges government contracts to be placed in this way. They lay down a formal procedure by which, for example, the offers have to be in sealed envelopes to be submitted by a fixed *closing date*, when the envelopes are opened in an official ceremony. The idea is to ensure fair play and avoid corruption.

This system was developed many years ago when most projects, however large, were fairly straightforward. That is still the case with standard civil engineering jobs, of the kind where the work is measured by quantity surveyors. Many of today's contracts are much more complex, which makes it more difficult.

PROBLEMS

I spent many years working in and running proposals departments, which is where a contractor's tender is produced. There I learned an important lesson, my first law of competitive tendering:

> *A contract that is let to the lowest bidder has gone to the contractor who made the biggest error in his proposal.*

The second law follows on from this and states:

> *Having won the contract, the lowest bidder will employ the lowest bidding suppliers and subcontractors in an attempt to reduce his direct costs.*

The corollary is that almost every major project is being built by a contractor who doesn't have enough money to do it properly. One of the reasons for this is the way in which tenders have to be submitted within a relatively short tender period, which means that the designers and cost estimators have to get it right the first time—there is no time available for a second try.

At best, things can go well, but that takes reasonable trust and goodwill on both sides and some spirit of achieving a common objective. It

Errors in proposals may be in the design, the cost estimate, or in the scope of supply. I well remember winning a contract by a margin of about 10% from a competitor. About a week into the contract period our contract manager asked me, politely, where the costs estimate was for the very complex control and telemetry system that the purchaser had specified. I had to admit that I hadn't read that page of the specification. The control system, in compliance with the purchaser's specification, added about 20% to the contract price, so I excused myself by pointing out that if I'd got the scope of supply right, our price would have been higher than our competitor's and we wouldn't have won the contract. It's a bit of a moot point whether we were better off winning it or losing it.

John Glenn, the first man to orbit the Earth on February 20, 1962, was asked what went through his mind while he was crouched in the rocket nose cone, awaiting blastoff. He replied: "I was thinking that the rocket had twenty thousand components, and each was made by the lowest bidder."

means that the personal relationship between the purchaser's *project manager* and the contractor's *contract manager* becomes all important, which reminds us that interpersonal relationships are every bit as important in engineering as technical matters.

At worst, this way of organizing a project is so adversarial that it satisfies no one—the contractor makes little profit or even a loss, and the purchaser gets an unsatisfactory end result.

The system only works perfectly when all the following conditions are met:

- The enquiry specification has to define unequivocally what is required.
- The contractor's proposal engineer has to design and cost the work accurately in the time available.
- The proposal writer has to define unequivocally what is being offered.
- The purchaser has to be able to assess which is really the best offer.

Of these, it is the last condition that can lead to the most difficult problems.

Government contracts are usually obliged to use this system, and many companies follow much the same pattern, though they are not legally obliged to do so. Some, such as the big oil companies, protect themselves against the worst disasters by having lists of approved contractors and won't consider anyone not on the list. These lists are limited to companies who are reasonably reliable and avoid awarding contracts that are too big for a small contractor by limiting the value for which the contractor may bid.

For some large international jobs, contractors have to present their credentials and/or buy the contract documents for a sum that is large enough to discourage frivolous offers.

The procedure traditionally goes back to simple civil engineering projects, with detailed specifications and no complications. Where that is the case, the purchaser looks at the offers, checks whether they comply with the specification, satisfies himself that the cheapest offer is by a contractor who can be trusted to perform the work, and places the order with him. Simple.

NOT SO SIMPLE

Most large contracts nowadays are more complex. When the project is first given the go ahead, the purchaser (or his consultant—and I shan't go on repeating this) may

The contractor's representative at meetings may be a contract manager, or a proposals engineer , or some such title. In a small company he may even be the CEO. The function he performs at the meeting, however, is that of technical salesman. Engineers (who can be terrible snobs) may look down on "salesmen," but this is a very skilled job. It not only means a good grasp of the technology but also a talent for inspiring confidence in this skill. He also needs to understand all the financial implications the contract would have for both parties.

invite one or more potential contractors to discuss it with him. These pre-tender discussions help the purchaser and contractor to understand one another and what the project is all about. The Purchaser may well get some good ideas from the contractor before he writes the specification.

The bit about opening sealed envelopes makes it all look absolutely fair, but reality may be rather different. When the offers have been examined, the purchaser will hold "post-tender meetings" to discuss it—possibly only with the successful contractor to resolve any difficulties or uncertainties, or with several contractors who are still in the running to decide which he likes best. He may even suggest to them to make some change to their offer and then retender. Plenty of opportunity for shady dealings there!

Sometimes there is conflict within the purchaser's camp: Maybe the technologists know which contractor they want to work with, even if he isn't the cheapest. This may be because they have worked happily together before, or maybe the cheapest contractor looks unreliable, or maybe the technology on which the cheaper offer is based looks doubtful. Or perhaps the people who are to operate the new facility want a dearer offer to win because it will be easier and more reliable to operate. The purchaser's purchasing officer (or whatever his title may be) will insist on the cheapest, regardless. "What's wrong with the lowest offer?" he asks. "Well, er…" it may be quite difficult for the engineers to put their preference into convincing numbers. "Has the

One of my unhappiest memories: A water company issued an enquiry for a plant to remove nitrate from drinking water supply. It was to use a novel process on which I had done a lot of research. Thanks to that I could put forward an alternative that would be so much cheaper and easier to operate that I thought we couldn't fail to get the contract. I hadn't reckoned with the post-tender meeting at which the purchaser gave a locally based competitor the details of my alternative (sadly none of which was patentable) and suggested he retender on that. Surprise! His offer was cheaper and he got the contract. Pretty dirty play, I thought, but it wasn't actually illegal.

A water treatment plant for a nuclear power station was to be fed with city water supplied through an old cast-iron ring main. This was to be dedicated wholly to supplying the power station, for which purpose the direction of flow would be reversed. The contract said we must satisfy ourselves as to the quality of the water, but as it was a public supply of drinking water we felt safe.

The plant started, and the first process unit clogged solid—not just once but again and again. The purchaser refused to pay. A close look at the rubbish that clogged it showed it contained (among others) freshwater shrimp. Quality standards for potable water don't allow shrimp, so we claimed we had been misled and they had to concede that we had been.

What had happened was that over many decades a great fauna and flora had grown on the old cast-iron main's walls. When the direction of flow was reversed they were facing the wrong way and came off.

contractor given us a guarantee?" "Well, er, yes." "So what's the argument about? We'll give him the contract."

GUARANTEES

Contracts often include some kind of performance guarantee—I used to say I could run a coach and four horses through any of them. When a project fails, the reasons are usually so various and so uncertain that it is impossible to pin responsibility on any one cause or even on one of the two sides in the argument. So it is usually difficult to make a guarantee stick.

Of course, if the specification in the original enquiry is wrong, the whole thing is doomed from the start. That is not the contractor's concern. Once he has the contract, his objective is to meet its terms as cheaply as possible. Whether the result actually satisfies the purchaser is not material—unless there are more projects on the way and the contractor wants to stay popular and get these other contracts. Even then he will only exert himself to avoid looking as if the fault is his.

In my specialized industry it often happens that the enquiry specification is wrong because of the purchaser's ignorance. At the pre-tender stage a good technical salesman will try to put him right, but that doesn't often work. Then the contractors can either give up any hope of getting the contract or offer a product that they know isn't up to the job. If their tenders are worded carefully enough, they're legally in the clear.

A contract for a UK power station at Warrington (Paul Simon wrote *Homeward Bound* on the railway station there) presented a novel process problem. I did some work on it, got it wrong, and our plant failed its acceptance test. It turned out to be the costliest mistake I ever made. The Purchaser was about to place two more large orders, so our keen sales director (see Chapter 7) was moved to offer a remedial plant free of charge. The only available site for installation was downwind of the cooling towers. "What does that mean?" I asked. "There'll be a constant drizzle of cooling water." "And what do you use for cooling?" I asked. "It's River Mersey water, unfiltered." they replied. I thought about this for a bit and said, "You mean: *The quality of Mersey is not strained/ It droppeth as the gentle rain from heaven.*"

Nobody laughed, and I still wonder whether they had no sense of humor, or just didn't know their Shakespeare.

So this way of placing orders can easily go wrong. With more and more specialized technology featuring in many projects, few purchasers have in-house the detailed knowledge needed to write an enquiry specification. Nor, when the tenders are examined, can they judge just what the contractors are offering and which really is the best option. The remedy is either to engage a consultant who does, or to tie the contractor into a performance guarantee which puts the responsibility firmly on him—but that's not easy.

To repeat: The main problem with this way of placing contracts is that contractor and Purchaser do not have a common objective. Sometimes, indeed, the proceedings become really adversarial. No sensible system would let contractors build projects that they know are unsound. No sensible system would encourage contractors to cut every corner they think they can get away with. (Even in straightforward civil engineering contracts there have been many cases where it transpires that the contractor had increased his profits by using a leaner concrete mix than that specified.) Hence the search for better ways of arranging things.

A BETTER WAY?

Placing contracts *by negotiation* means that instead of asking contractors to send in competitive bids, the purchaser discusses the project with potential contractors and selects the one who seems the best. The purchaser and the appointed contractor then go through the project in detail, agreeing on the specification and pricing its components as they go, to build up a specification and price that are the basis of the order.

The reimbursable contract (see Chapter 17) is a halfway house that combines competitive tendering with a negotiated contract price.

If this type of approach is to work, the purchaser must have the expertise and the resources to deal with all the technical details—without this resource the negotiation would be pointless. The system should result in a better contract (for both parties) than the adversarial system does, but it only works because a major source of failure has been removed. And once the contract is signed, the two systems are the same.

> I visited Japan in 1965 when we had just started to work with Japanese main contractors who favored negotiated contracts. I was struck by the number of engineers they could fling into solving any problem that might arise. It made certain the problem was solved, though their solution might be labored rather than elegant. I became friendly with my opposite number, who had just worked very long hours with no hope of any reward—in those days promotion was strictly by seniority. What motivates you, then? I asked, and he replied "I am an engineer. Engineering is an honorable profession."

SUMMARY

- **In competitive tendering, offers have to be prepared and priced quickly—mistakes can and do happen.**
- **The contractor will offer something to satisfy the enquiry specification, whether that specification is technically sound or not.**
- **Many purchasers don't have the experience or expertise to know which offer is the best: Choosing the lowest priced tender can be quite wrong.**

Chapter 16

How Things Can Go Wrong—3

I got this example by working as an expert witness in a large lawsuit in London's High Court. Expert witnessing is a great experience. For one thing, it is splendidly well paid, like everything to do with the law, of course, and you will discover that top lawyers are great people. It cured me of an engineers' prejudice against that profession.

The background was this: a large chemical works needed a constant and reliable supply of steam for its processes. The client was a joint venture company that had been formed in order to build a gas-fired power station to supply this steam. The plan was to generate steam at high pressure and use it to generate power before going over the fence to the chemical works.

If the steam supply were to fail, however, the chemical works would have to shut down. If the reaction tanks and pipelines are allowed to cool, they set solid with polymer, and it can take weeks to clean them out. Reliability is obviously important.

Expert witnesses are employed by solicitors acting for companies who are engaged in civil lawsuits that involve specialist knowledge. Legal procedures vary in different countries and change from time to time. This case was conducted under the traditional UK system, but an alternative method using a single independent expert witness is now an option, although it doesn't appear to have made much headway.

In the conventional system each side's solicitors engage an expert in the field in question. In complex problems there might be several pairs of experts, each dealing with one aspect. The expert must be totally truthful but will naturally stress whatever favors his side of the argument. The two experts each write reports that they exchange before they meet in the hope of agreeing on as many points as they can.

Sometimes there is no agreement at all, but I once had my opposite number (an old friend) come to the meeting, in total agreement with my report. "There was really no need for us to meet", he said, "but I wanted an outing and to say hello to you. This project was a mega-cock-up, and that's it. Let's go and have a beer."

Engineering Money: Financial Fundamentals for Engineers, by Richard Hill and George Solt
Copyright © 2010 John Wiley & Sons, Inc.

The client engaged a well-known consulting company to write the spec for a turnkey contract to build this power station. The spec was generally functional, but it did include detail for some of the major items. Several contractors submitted offers for the job.

The consultants found that the cheapest offer did not measure up to the standard required by the spec and was thrown out on that account. It then transpired that the contractors' prices for all the offers came out at more than the client's budget, and that included even the cheapest one from a major U.S. contractor that had been thrown out as noncompliant.

The client's managing director and the CEO of the contractor who had made the cheapest offer had worked together before and were on good terms. The two sat together and went over the detailed spec that was the basis of the contractor's offer. They cut out everything not immediately vital to the operation of the power station, like standby equipment. When they had finished, they had reduced the contractor's price by 7%, and brought it within budget.

> There is an important difference between procedures in different countries. The High Court in England has a "TCC" (Technology and Construction) branch, where the judges are trained in some branch of science or technology. This is not the case in the United States, so that there all the arguments have to be presented in a way that a lay judge can understand. The same applies to most European countries.

The consultant, sensibly, would have nothing to do with this, and withdrew. Client and contractor were on their own.

The contractor was given the order, based on the copy of the specification that the two of them had slashed. It was a photocopy of the contractor's original with the crossings-out and notes that the two of them had made on it. Some of these were not too clear, and one or two were actually contradictory.

It was not a good basis for a multimillion dollar project because:

- The original offer had been turned down by the consultant as "noncompliant."
- They had taken a great deal out of that, presumably making it even more "noncompliant."
- The resulting document was neither clear nor consistent.

Not surprisingly, a great deal went wrong. By the time the station

> The actual chain of events went like this: The contractors had arranged a performance guarantee with their bank. This means that the bank undertakes to pay a sum of money to the client if he claims the contractor has not performed the contract properly, and debits the contractor for it. The bank must pay up on demand, without judging whether the demand is justified or not. It is then up to the contractor to bring an action to recover the money. So the contractor brought the action because the client effectively forced him to do it.

started up, the client's original MD had left. His replacement found things unsatisfactory (which indeed they were) and—unwisely, as it turned out—instigated a lawsuit against the contractor.

One of the more important failures was that the boiler feed water plant (my specialized field) was too small, had no backup capability when one or other of its components had to be taken out of service for maintenance, and the treated-water storage tank was too small to maintain the supply for such periods.

There were other problems such as, for example, it was claimed that the waste chemicals from the water treatment plant created a dangerous effluent. Quite separately, it seems that the cooling tower wasn't up to its duty—other expert witnesses were engaged for that.

The boiler feedwater plant was designed and built (very cheaply, no doubt) by a Canadian subcontractor. We now had the client (and his site) in England, the main contractor in California, and the subcontractor in Canada. Communications got fouled up, and when there was trouble on the site, there was inevitable delay in sending out people. It was a recipe for confusion and delay.

> The water treatment plant produced waste streams of acid and caustic that were to neutralize one another in an effluent tank before discharge. The tank was actually too small, and there were times when its contents were quite acidic. The client got the Health and Safety Executive to make an issue of the resulting danger. A colleague of mine was engaged to deal with this aspect, and it proved his finest hour. He produced a bottle of dilute sulfuric acid in court, poured it into a beaker, and stuck his index finger into it. "How long would your Honor like me to keep my finger in it?" he asked. The judge smiled, told him to take his finger out, and we heard no more about the very dangerous effluent.

Still worse, the plant had been designed using a technique that my company had abandoned 20 years before because of its high chemical consumption. I was working on the contractor's side and would say to the solicitors, "This is the lousiest plant I have ever seen." They would reply, "Yes, but does it conform to the contractor's offer?" Actually it did because the specifica-

> *"A lawyer will do anything to win a case, sometimes he will even tell the truth."*
> —Patrick Murray, American comic

tion said nothing about its chemical consumption, so I had to say, "yes, I suppose it does." "Well, in that case be so kind as to keep your opinions to yourself. All we have to establish is that the plant conforms to the contractor's offer." We won, but in my view both sides emerged pretty badly from the shemozzle.

SUMMARY

- **A contract is based on a specification that has been agreed upon between purchaser and contractor.**
- **Once a contract has been formed the contractor's duty in law is to meet that specification.**
- **If the specification is incorrect—not what the purchaser wanted—then it is the purchaser's responsibility not the contractors.**
- **Only lawyers get rich on specifications that are not clear.**

Chapter 17

Other Types of Contracts

TRADITIONAL APPROACH

Traditionally, contracts are generally of three types:

- *Supply only* in which the contractor has to supply and deliver certain specified goods—for example, a factory-built skid-mounted pump set—to a specified delivery point at a specified time.
- *Supply and install*, where the contractor is required not only to supply the specified goods—for example, an air-handling unit for an air-conditioning system or underground pipes and fittings—but to install them and, often, commission them.
- *Build only* in which the contractor is required to construct exactly what is specified in the contract.

Selection of the contractor is by competitive tendering (see Chapter 15). The lowest bidder wins the contract even if (or quite often because) he's made a mistake in his tender price. The conditions of contract generally promote a confrontational or adversarial relationship between the purchaser and contractor with the result that neither party is really happy with the result (see Chapter 11). To overcome some of the intrinsic problems of competitive tendering, various approaches have been developed to find a way of giving both purchaser and contractor the common objective of producing a result that represents the best solution at the lowest price.

REIMBURSABLE CONTRACTS

Since the mid-1980s, particularly in the process industries, there has been a move toward *reimbursable contracts*, in which the contract price is not fixed, but the contractor is reimbursed by the purchaser for materials and labor as the work is completed.

Engineering Money: Financial Fundamentals for Engineers, by Richard Hill and George Solt
Copyright © 2010 John Wiley & Sons, Inc.

The way these contracts are awarded is more complicated than conventional competitive tendering. The purchaser selects a short list of contractors whom he thinks capable. He interviews these *prequalified* contractors and selects one. He than issues a performance specification to the selected Contractor setting out what the project has to achieve.

The contractor is free to propose any solution he likes. He may be a specialist contractor who has a particular technology that he wants to use, or he may be a main contractor who can choose from a wide variety of technologies on the market and employ specialist subcontractors to supply them. Because of the range of technical skills needed in this type of contract it is quite common for a consortium to be set up to tender for very large or complex contracts. This is a separate company specifically formed for the contract, in which a number of contractors have a share.

Once the contractor has been appointed, the contract usually proceeds in two stages. In the first stage the contractor prepares a *front-end engineering package* that has enough design information to produce a more accurate cost estimate. The design is agreed upon with the purchaser and the cost estimate becomes a *target cost* for the second stage of the contract.

Let's take an example from an industry I'm familiar with. Suppose a water supplier wants to build a new water works. The purchaser (the water company) issues an enquiry document describing the site, giving information about water quality and the quantity of water he wants to supply, the ground conditions and similar information. It will also give the selection criteria—this may be capital cost, life-cycle cost, robustness of plant, footprint, environmental impact, and so on.

Each of the prospective contractors prepares a proposal describing how he will do the job—what water treatment processes he will use, what the buildings will look like and estimated capital and operating costs for his proposed water works. The contractor can propose whatever he thinks is likely to be the best solution: maybe a combination of traditional processes such as clarification, filtration, and disinfection or, perhaps, a high-tech solution such as reverse osmosis. The purchaser selects whichever contractor he thinks has the best solution based on the selection criteria.

The second stage of the contract is pretty much like any other design-and-build contract except that the purchaser is much more closely involved in all the decision making. The target cost estimate identifies the cost of each direct cost item—materials, subcontracts, and engineering, procurement, and contract management person-hours at the contractor's cost center rates. It will also identify an agreed upon markup (that is a percentage addition to direct costs) to cover the contractor's overhead. This might typically be around 15%. Thus each direct cost item has 15% added to it as contribution. As the contract proceeds, the contractor expends person-hours, buys materials, and places subcontracts. Each item of expenditure is accounted for in detailed contract management accounts that both the contractor and the purchaser can review at any time. This is called *open-book* accounting and both parties know exactly what money is being spent and how much profit the contractor is making—a situation with which many contractors are not comfortable. Progress is monitored weekly and any variations from the target cost have to be agreed upon with the purchaser.

Design and construction problems occur in most contracts. In conventional contracts they are usually a problem for the contractor who would then try to claim the cost back from the purchaser as an extra. The purchaser would try to resist any additional cost. In reimbursable contracts the purchaser and contractor are part of the same team and have to work together to solve the problem.

Two things are obvious here. First, it is essential that both purchaser and contractor have technical expertise. This is not a matter of placing a contract and leaving the contractor to get on with it. The purchaser is involved as part of the team and he (or his consultant) must have at least as good an understanding of the technical aspects of the project as does the contractor, so that meaningful discussions can take place prior to technical decisions being made. Second, it requires a degree of trust between contractor and purchaser, which is still a novel notion in the construction industry, at least in the UK.

> I was involved in a landfill leachate treatment plant contract where the ground conditions were thought to be good. During the contract, it became obvious that the site would need to be piled—an expensive option. In a traditional contract the contractor would have had to pile the site and hope to claim some of the costs back from the purchaser as an extra. But in this reimbursable contract the contractor and purchaser had to work together as part of the same team to find an acceptable solution to the problem. The solution involved some redesign to spread the load and a more expensive robust foundation, but one that didn't need piling.

As we have seen, special conditions of contract are applicable to these reimbursable contracts, the most widely used being the Institution of Chemical Engineers' *Green Book Conditions* (see Chapter 11).

A major advantage of this type of contract is that it can shorten construction time by allowing *fast track* projects, where the design is not finalized at contract award. When the design changes, the contractor is paid for the design time at the agreed rate.

A DIFFERENT APPROACH

Over the last decade different types of contractual approach have been tried including:

- Design and build (D&B)
- Design–build–finance (DBF)
- Design–build–operate (DBO)
- Design–build–finance–operate (DBFO)

- Build–own–operate (BOO)
- Build–own–operate–transfer (BOOT)

In D&B contracts the contractor is responsible for the design as well as for the construction. This relieves the purchaser of the responsibility for designing the project (or paying a consultant to do so) but means that the tenders received may all be based on different designs, and this makes tender adjudication a much more complicated procedure. On the other hand it gives the contractor the opportunity to offer an innovative design that he couldn't have offered under the traditional build-only contract.

In DBF contracts the contractor is again responsible for the design but also provides financing for the project. This means that the purchaser doesn't have to pay interest on a bank loan (although the cost of the contractor's finance will be included in the contract price). This can be useful if the purchaser doesn't want interest payments to appear on his accounts, or if the contractor can borrow money more easily or at a preferential interest rate.

Design–build–operate contracts are a bit more complicated. Here the contractor not only designs and builds the project but also operates it. In the case of a road the contractor will have a maintenance contract for a period of, typically, 25 years. If it's something like an effluent treatment plant, the contractor will provide all the necessary operating staff and support services. The benefit to the purchaser is that he doesn't have to worry about providing specialist operating services, and a contractor with a number of operating contracts should be able to reduce costs by centralizing common services.

Design–build–finance–operate contracts are the same as DBO contracts but the contractor also provides the finance.

Essentially, BOO contracts are the same as DBFO contracts, but in a BOO scheme the contractor retains ownership of the project and would write off the value of the capital cost in his operating charges over the contract period. This means that the purchaser effectively buys whatever the project produces. In a word, it's outsourcing. A fairly common example is the setting up of combined heat and power (CHP) projects. Here a consumer of steam and electricity—say an oil refinery—scraps its boiler house and buys both steam and electricity from the CHP plant operator at an agreed price on a long-term contract (see Chapter 16).

> In the water treatment business, this approach was first used in Silicon Valley in California. Making computer chips needs ultra-pure water—water so pure you can't measure the impurity levels in it. The chip makers have their own highly specialized technology. They don't know, and don't want to learn, about designing and operating plants for making ultra-pure water. Instead they give a specialist contractor a bit of land to build a plant by the side of their works, and buy the ultra-pure water on a long-term contract.

Table 17.1   Responsibilities

Type	Design	Construction	Finance	Operation	Ownership
Build only	Purchaser	Contractor	Purchaser	Purchaser	Purchaser
D&B	Contractor	Contractor	Purchaser	Purchaser	Purchaser
DBF	Contractor	Contractor	Contractor	Purchaser	Purchaser
DBO	Contractor	Contractor	Purchaser	Contractor	Purchaser
DBFO	Contractor	Contractor	Contractor	Contractor	Purchaser
BOO	Contractor	Contractor	Contractor	Contractor	Contractor
BOOT	Contractor	Contractor	Contractor	Contractor	Contractor

The BOOT schemes are essentially the same but, at the end of the contract period, the whole thing becomes the property of the purchaser. There are a number of variations on this system, but the essence is the same.

Unless there is a reason that the purchaser does not wish the fixed-asset value to appear on his balance sheet, the decision between the various financing options will be on cost. All other things being equal, this rests on the cost of capital. In each case the plant operating costs should be the same, and the only difference will be the cost of borrowing. If financing is placed with the design-and-build contractor, he will normally employ a leasing company and offer either a straightforward lease (BOO) or a lease–purchase scheme (BOOT) under which title would revert to the purchaser on termination.

The responsibilities of the two parties are summarized in Table 17.1.

The last four of these share a mold-breaking characteristic: The contractor does more than just provide buildings and hardware—he has to operate the plant: That is, he has to provide a service over a long period. This has far-reaching implications.

The skills needed for plant operation are different from those needed for design and construction; thus, the consortium or joint venture company (JVC) approach is widely used in these various types of contracts (see Chapter 12). Here the construction contractor and operating company are shareholders in the JVC, so they share in the risk and profit with the shareholding reflecting the risk that each party is taking and the scope of their input to the contract.

Under a conventional design-and-build contract, which is awarded on capital cost, the contractor will generally engineer the plant to a lowest capital cost, which will usually mean higher operating costs for the client. For example, typically pipe velocities will be high, giving low-cost pipes but high head loss and, therefore, higher power consumption. This option may tempt the purchaser with its low initial cost, but cost him dear in operation.

A DBO scheme should, in theory, lead to the highest capital cost for the purchaser since the contractor will seek to maximize the capital cost (which will be paid by the purchaser) and minimize the operating costs for which the contractor will be responsible.

As we have seen, conventional contractors only have their working capital; thus, they can't raise the money necessary to finance large projects without help. The way they overcome this is for the JVC to include a finance company in its members.

The design, construction, and long-term operation of the project are all in the same hands so, in theory, this should mean that the objective is the minimum life-cycle cost of the project. This being the case, there is no point in skimping at either the design or construction stages, if the effect will be increased costs of operating or maintaining the product. The problem is that there is a conflict between the interests of the construction contractor and the JVC in which he is a partner. The JVC's interest is a long-term one, but the construction contractor is focused on this year's accounts. Consequently, the construction contractor still has an incentive to minimize his costs and maximize his current profits; thus, as in conventional contracts, he will aim to just meet the specification.

Traditional contractors who have moved into DBFO projects find that the entire character of the company changes. It now owns fixed capital, which gives it stability and which banks consider as security for loans. It has a steady income for many years to come, and it no longer survives by going from contract to contract. It no longer provides a finished product, but a long-term service. Bankers are generally pleased at the stability and the long-term income which DBFO projects bring.

> A water company needed a sewage sludge processing plant and outsourced it as a DBFO scheme. The JVC consisted of a construction contractor who would build it, another water company who would operate it, and a bank who financed it. There was an expensive pressure vessel implosion. The operating company blamed the construction contractor on the grounds of bad design and the contractor blamed the operating company. As a result, two partners in the JVC (or rather their insurance companies) finished up suing each other (see Chapter 12).

PRIVATE FINANCE INITIATIVE

Monetarist governments worldwide are increasingly turning to private finance initiative (PFI) schemes under which public services such as hospitals, prisons, universities, and roads are designed, built, financed, and operated by private companies. The operating company is paid on the basis of the number of units—patients, prisoners, students or vehicles—using the facility. PFI schemes are generally very large and JVCs are invariably set up to carry out these projects.

> A typical UK private finance inifiative contract was for a stretch of highway—the traffic on it was monitored and the operating company was paid so much per vehicle. This meant something of a gamble because the income couldn't be predicted with great confidence—it would depend on the state of business in general and local conditions in particular. The contract also specified that if any lane of the highway were closed for maintenance, the payment would be reduced—an excellent incentive for sound construction.

The bad news is that many PFI contracts have gone shockingly wrong, and some well-publicized examples have been very costly for the taxpayer. The ones where the JVC is in trouble are less well publicized. It is also true that PFI is a radical new idea that still needs to be run-in properly—at the time of this writing, there can scarcely be any PFI schemes that have already run their full contracted course. With more experience it should be possible to do better.

The reason for the failure of at least some of these PFI contracts is the fact that governments in general seem incapable of organizing good contracts. UK defense contracts regularly go 200–300% over budget, and various IT projects have even worse records.

Speaking personally and as an engineer, I do hope that something along the lines of DBFO will emerge as a practical method of commissioning projects. It holds out the promise of better engineering. By definition, that means a better outcome for everyone.

SUMMARY

- **The traditional competitive tendering method of commissioning projects has often given bad results.**
- **The last 20 or so years have seen a variety of efforts to improve matters by different forms of contract.**
- **Some of these turn the contractor from a provider of hardware into a service provider, and that should, at least in theory, make for better design and construction.**
- **No clear winner has yet emerged.**

Chapter 18

Terms of Payment

WHAT ARE TERMS OF PAYMENT?

Purchasers understand that contractors need to finance the contract and will be aiming for a positive cash flow. The conditions of contract will, therefore, include a set of agreed terms of payment that detail how the contract price will be paid to the Contractor.

Contractors do not generally

> Paying interest on the money is valid even if the contractor has enough cash in the bank to be able to fund the work without borrowing. If the cash is not needed to fund the contract, the contractor could have put it out on loan, to a local authority or building society, for example, to earn interest. Either way interest is lost.

have the working capital needed to finance big contracts but, sooner or later, the purchaser will have to raise long-term capital for it anyway. It would be more expensive for a contractor to have to raise this for the short duration of the contract, even if the bank would let him have a loan. And he would have to include the cost of the finance in the contract price (see Chapter 14). It is usually cheaper for the purchaser to raise the capital himself and pay the contractor as work proceeds. Aside from the issue of finance costs, there is the question of ownership, as we shall see later.

Therefore, except for very small contracts, it is normal for the purchaser to pay the contractor in stages as work proceeds. The purchaser, of course, has similar problems of interest and cash flow. There is a direct conflict between the purchaser, who wants to pay as little and as late as possible, and the contractor, who wants the opposite. Setting acceptable terms of payment is, therefore, important to both parties. Construction contracts usually allow for payments to be made at monthly intervals against measured work completed or other *milestones* that are defined in the conditions of contract. The sort of things that might count as milestones include completion of certain specified drawings, completion of foundations, installation of mechanical and electrical equipment, and so on. Payments such as this are called *stage payments*.

Engineering Money: Financial Fundamentals for Engineers, by Richard Hill and George Solt
Copyright © 2010 John Wiley & Sons, Inc.

OWNERSHIP

In mechanical and electrical (M&E) and process plant contracts, where a lot of the fabrication work is carried out off-site, the contractor will probably have to spend a lot of money before anything is actually delivered to the site. A typical contract might include the supply, delivery, installation, and commissioning of steel pressure vessels—perhaps a boiler system. It is quite usual to provide a payment to the contractor when construction materials (typically steel plate or pipework) are delivered to a fabricator's works, provided that the materials are identified as the purchaser's property. This means that the purchaser has a *lien* (a right of ownership) on the materials—that is, he will be able to claim them should the fabricator go into receivership before completion of the contract. Similar stage payments may

> The *title*, that is, the ownership of the goods, normally passes to the purchaser only when money has been paid. The *risk* may or may not pass at the same time. This means that although the purchaser may own the pressure vessels before they are delivered, the safekeeping (in practice the insurance) of them may still be the contractor's responsibility. If there is a fire at the workshop and the vessels are damaged, making good is likely to be the fabricator's responsibility rather than the purchaser's. It is, therefore, important to distinguish between title and risk in goods and at what stage each passes from one party to another.

be made on completion of the fabrication, on delivery to site, on completion of installation, and on completion of commissioning.

In design-and-build contracts it is usual to allow stage payments against completion of all or parts of the design, typically on presentation of drawings and calculations. The copyright in the drawings will usually pass to the purchaser on payment, but the laws relating to intellectual property are beyond the scope of this book.

DELIVERY

It might be useful to introduce a few terms that you might come across in export contracts. In home construction contracts the contractor is normally involved in supply, delivery, and installation of equipment; so the contractor is responsible for getting the equipment to the site, unloading it from the truck, storing it until it's needed, moving it into position, and fixing it (although there may be

> I worked for a UK company that routinely built skid-mounted water and wastewater treatment plants for export contracts. A typical export contract would be the supply of a skid-mounted produced water separator for separating crude oil from produced water on an offshore platform. We would engineer and build the unit and ship it out to wherever the platform was being constructed, usually in Scandinavia.

some argument between the contractor and his supplier about who should be paying for the transport and craneage for offloading). Things are often less clear in export contracts. This is where we need to understand the various terms for delivery.

The commonest delivery terms are:

ex-works: The contractor has to make the goods available at his premises (that's the factory where the equipment was assembled). He is not responsible for loading the goods onto the purchaser's truck unless it's been agreed separately.

FAS: This is short for free alongside ship and means that the Contractor has to place the goods on the quay next to the ship and has to pay all the costs up to that point including any export clearance costs. Once the goods have been placed on the quay, they become the purchaser's property and he has to bear any risks.

FOB: This is short for free on board and here the contractor is responsible for placing the goods on board a ship—usually nominated in the contract—including paying for any documentation costs. Once the goods are on board the ship, the captain will issue a *bill of lading* as proof, and this usually has to be presented to the purchaser to get payment released. The risk passes to the purchaser as the goods cross the ship's rail.

C&F: means cost and freight and here the contractor is responsible for all the costs involved in getting the goods to a destination nominated in the contract. The risk passes to the purchaser as the goods cross the ship's rail at the port of shipment.

CIF: means cost, insurance, and freight, and is essentially the same as C&F except that the contractor has to pay the insurance premium for marine insurance against loss or damage in transit.

Other terms might include the contractor being required to pay local import duties and taxes at the destination.

In export contracts payment is often made by means of an *irrevocable letter of credit.* The Purchaser arranges with his bank to provide credit in a UK bank which is used to make payments to the Contractor on presentation of specified documents such as the Bill of Lading. The contractor deals only with the bank and the purchaser cannot refuse payment for any reason.

> The letter of credit system works well but must be handled carefully. The documents presented must have identical wording to the letter of credit. If a word is misspelled in the Letter of Credit it must be identically misspelled in the Bill of Lading otherwise the bank will not pay.

RETENTIONS

It is not unusual for the stage payments (see Chapter 9) to amount to only 90 or 95% of the contract price, the balance being retained by the Purchaser for some period (typically 12 months) after contract completion. This *retention* is intended to ensure that the contractor responds to his post-completion responsibilities, such as making good any latent defects.

Retentions obviously make a serious hole in a contractor's working capital (see Chapter 9). Suppose our company with the $40 million turnover has a 10% retention for one year on all of its contracts. That means that, on average, it has $4 million permanently held up in retentions. Given that the company needs about $8 million working capital, that's a major financial burden.

On large contracts retentions can tie up so much cash that it can cause major problems. In this case the contractor might arrange a bank guarantee as an alternative. The contractor is paid in full but gives the purchaser a bank guarantee for the sum of the original retention. If defects are found during the guarantee period and not rectified by the contractor, the purchaser is entitled to demand payment from the bank. The bank must pay up without question and, if the contractor thinks the purchaser is in the wrong, then they have to sue the purchaser for the money. The bank's charges make this too expensive for all but very large contracts.

EXTRAS

All contracts have variations during their course; work—maybe additional engineering or the supply of materials—that was not originally foreseen. The contractor will usually try to recover the costs of these variations by claiming them as an *extra to contract* from the purchaser. The purchaser will not always agree to pay, claiming that the work should have been included in the contractor's original estimate. This issue is the most common form of dispute in contracting and will often happen many times during the course of the contract. The important thing is not whether the work has been done or the contractual position with respect to fault, but whether there has been agreement from the purchaser that he will pay.

Suppose the contractor has to provide additional materials that he believes to be extra to contract, but the purchaser believes they should have been included. The additional costs appear on the contractor's balance sheet as a liability (because he has to pay them) and must remain as a liability until such time as the purchaser agrees to pay the extra, when they become an asset. This agreement may not be given until the completion of the contract; it may not be given at all and may become the subject of protracted litigation which may take several years.

The principle remains that the costs are a liability on the Contractor until the dispute is resolved. Overoptimistic project managers who allocate all claims for extras as assets before dispute resolution can cause major problems because, on the basis of the management accounts, those assets will appear on the balance sheet as work in progress and distort the true picture (see Chapter 4).

Some less than scrupulous contractors routinely bid low prices for contracts and employ very sharp quantity surveyors to negotiate extras through the contract period to make up the contribution that the original contract price lacked (see Chapter 11).

> "I knew we were going to have problems on this contract," the project manager told me. "The first thing the contractor (one of the top UK civil contractors) did was to install a site cabin with five Quantity Surveyors in it."

SUMMARY

- Except in the case of very small contracts, most contractors don't have the money needed to complete the contract without being paid part of the contract price as work proceeds.

- If these stage payments don't cover the expenditure on the contract, the contractor has to find the difference out of working capital. That affects cash flow.

- Retentions do that, too.

- Contractor and purchaser both want to nurse the cash flow and avoid borrowing.

- A compromise has to be found at the contract stage and set out in the terms of payment.

- Extras are a battlefield.

Chapter 19

How Things Can Go Wrong—4

The worst contractual disaster I ever got involved in was the construction of an effluent treatment plant for a chemical manufacturer. The client wanted to employ a respected firm of consultants to carry out pilot plant trials to establish the process requirements and the design parameters. The client also wanted the consultant to be involved right through the contract. An agreement was made between a major civil engineering contractor and the consultant that the contractor would take on the construction work but would employ the consultant to do the pilot plant trials and, importantly, to be the designer for the contract.

The client was happy with this arrangement and, in the mid-1990s, agreed to place a reimbursable contract with the civil contractor under which the pilot plant trials would be carried out, the design parameters established, and a cost estimate agreed as a "target cost." All was going well and the cost estimate was produced at £24 million. The Client did not have enough money and a number of changes were made in the scope to reduce the costs to about £15 million, at which price the contract was signed.

Things started well but the contract manager was a road builder with no experience of process plant contracts. The contractual arrangement between the consultant and the contractor did not specify who was to provide the all-important preliminary engineering documentation known as the *front-end engineering design* (FEED). The contract manager decided to start on the construction without the FEED, while the contractor and consultant argued about who should be doing what. Not a good start.

The consultant was accustomed to working as a client's representative, that is, overseeing the work of a contractor. In cultural terms the consultant's staff looked down on contractors as intellectually inferior (see Chapter 23). But this job was different because, in contractual terms, the consultant was a design subcontractor to the civil contractor. It was not going to be an easy relationship.

Engineering Money: Financial Fundamentals for Engineers, by Richard Hill and George Solt
Copyright © 2010 John Wiley & Sons, Inc.

It got worse. As the contract proceeded, the project manager invoked safety as an argument for some more changes to the scope, this time reinstating many of the things that had been removed to reduce the cost. When the contract manager pointed out that the project manager himself had asked for these items to be deleted, the project manager's response was that they were needed on the grounds of safety. Without them the plant would not be safe and the contractor's design was, therefore, flawed and the various items would have to be supplied *at the contractor's cost*. The disputes continued and the program slipped. By now relations between the project manager and the contract manager had deteriorated to the point that the contract manager had to be replaced.

His successor was a more experienced man but he misread the situation. In order to placate the project manager he agreed to most of the demands in the expectation that he would be able to renegotiate costs, which had by now risen to £25 million, later once the contract was back on program. He was wrong. The project manager was nearing retirement age: This was to be his last project and he was going to complete it within the £15 million estimate. Cost negotiations failed.

Sadly there was further bad news. The consultant, it turned out, had got his design badly wrong, and many of the plant items did not perform to specification. This caused further friction and, in the eyes of the contractor's engineers, confirmed the opinion that the consultant's engineers were too academic with no practical experience.

I had done subcontract work on other projects for both the consultant and the contractor during this time, so knew both teams and was aware of the contract's problems. Toward the end of the contract I was employed by the contractor to help tie up some loose ends. On one occasion I met the contractor's divisional CEO in the kitchen of the site cabin and he asked me what had gone wrong. I summarized my opinion as follows: "Your contract management has been incompetent, the consultant's design was wrong, and the project manager is the most obnoxious, unscrupulous engineer I have ever had to deal with. It was never going to be a success."

On completion of the contract, the contractor employed a firm of quantity surveyors to draw up claims for most of the £10 million overspent. Since they hadn't been involved in the project construction, they were depending on the contractor's site staff and the consultant's engineers to provide much of the information. I was asked to meet the quantity surveyors to add what I could to their knowledge of the project. There were six of them in a hired office about $10\,m^2$. The walls were floor-to-ceiling bookshelves packed with the files on the project. It became quite clear that, while their first task was to prepare the claim against the client, they were also building up a case against the consultant in case the claim failed.

Almost 10 years after the project's inception, the contractor started litigation procedures. The client not only threw out the claim but issued a counterclaim for as much again. Inevitably the contractor sued both the client and the consultant. All three sides employed solicitors, barristers, and other consultants as expert witnesses. By this time nearly all the people involved in the project had moved on to new companies, retired, or, in one or two

> *"A jury consists of twelve people chosen to decide who has a better lawyer."*
> —Robert Frost

cases, died. The arguments put forward were argued by people who had had no involvement whatever in the project. And when all the arguing was over and the accounts were done, the only people who had really benefited from the project were the lawyers.

SUMMARY

- **This contractual fiasco broke all the rules of good contractual practice.**
- **Perhaps the most important lesson you can learn from it is that contractual problems rarely have anything to do with technology.**
- **Mostly they are interpersonal problems between the individuals involved.**
- **And, of course, money.**

Chapter 20

Planning Contract Execution

WHAT NEXT?

Our contract is signed and the rules under which it is going to be executed are fixed. At last (after 19 chapters) we can get to consider the business of actually doing it. From now on, in a conventional contract, the contractor has two main objectives:

- To get by his contractual undertaking while spending as little as possible

- To get the greatest benefit (or the least damage) to his cash flow

"Rules" here has the same meaning as in sports, where your game plan aims to extract the greatest possible advantage from the rules. Over the last decades soccer fans haven't had a lot of changes to the rules, so they will be less aware of this than followers of the Rugby Union. Constant fiddling with the rules of rugby in the last 30 years has brought about huge changes in how rugby is played. And the interpretation of the rules isn't quite the same in Europe as in the Antipodes, so players who move from one to another need to adapt. The same adaptability is needed for performing a contract. No wonder management games are much used in management training courses.

The basis of the contract is the specification, which was agreed to when the contract was signed. It doesn't matter now whether this was part of the purchaser's enquiry or of the contractor's offer—anyway, some details may well have been modified during the contract discussions.

Contracting is a trade that has its own skills, regardless of the technology involved, whether constructing an oil refinery, just doing the electrical wiring for it, or for building a bridge. Of course, any such project requires expertise in its particular technology, but here we are dealing with the skills that must be used in the execution of all of them—the professional skills used in executing projects.

Time is the essential component in this. Time is the one irreplaceable commodity we have, but there are sometimes ways of buying it with money. Planning is the art of organizing time and identifying where changes could help to speed up the work. In our context that means providing the labor, materials, and services for the job at the right time and place, monitoring the work as it proceeds, and taking corrective action as soon as it is needed.

Any part-finished project is a waste of money for two reasons: loss of benefit from the completed work and loss in the form of interest payments on the money that has been spent. Suppose, for example, that Wal-Mart acquires a store. Management expects to spend quite a lot of money refitting it, but that expense is small compared to the profit it hopes to make when it opens. Every day's delay in opening costs Wal-Mart a day's takings. Moreover, the money paid for the site and the money so far spent on refitting is all costing interest. This is why contracts are often drawn up with bonus payments to the contractor for early completion and large liquidated damages for lateness.

THE PLAN

Making a plan consists of reducing the job to its component elements or *activities* and then fitting them into a timetable. For each activity we need to know:

- How long is it going to take?
- Is its start dependent on some other job having been completed first?

And then, when we come to arrange the elements into a timetable:

- What materials, services, and the like will each element need?
- How long will it take to procure those and when are they available?

With this information (and other minor items) it is possible to set up a plan. While the purchaser's objective here will be to complete the whole job as quickly as possible, the contractor's normal objective is to reach his payment points as quickly as possible. These two objectives are usually identical, but not always. Either way, we want to shorten the time taken for some or all of the job. To do that we must first identify the sequences of elements that have to be carried out one after another.

When building a house, for example, building the walls needs the foundation finished, and putting up the roof needs the walls to be finished, but digging the rain water soakaway can be done at any time. When we have identified the order in which the elements must follow one another, and how long each one of them takes, we can identify which of these sequences takes the longest time.

Critical path analysis was developed in the United States as a formal technique during World War II in order to minimize the time it took to refit aircraft carriers. As long as the carriers were in dock not doing their job of protecting convoys against German submarines, ships would be torpedoed and their crews would drown. This wasn't about saving time—it was about saving lives.

Once we have identified this controlling sequence, we have the *critical path*. Shortening anything that does not feature on it can't shorten the whole job. There is a formal way of doing this exercise called *critical path analysis* (CPA), which is explained briefly in Appendix 2.

Having identified the activities on the critical path, you can then consider what you might change in one or more of the components on it in order to shorten the overall time. There's no point in doing anything with an activity that isn't in the controlling sequence—it isn't going to help.

It is a universal fact that in any job there is a controlling sequence that is worth improving, which is the basis of CPA. It is true for any job that has a number of activities. It should never be ignored when putting together a plan, even if it isn't worth doing a formal CPA on it.

However formally or informally the plan is laid out, the best way is to start the process with the finishing point and work backwards. At every stage in the process you then have to ask "What needs to be finished and available before I can start on this element?"

The next stage is to look at the activities on the critical path and see whether it is possible to shorten the time each of them takes. If that is possible, it may be that another path then emerges as the longest— criticality has been transferred. In a really complex job such as building an oil refinery, the network is so complicated that this is best done on a computer. There is a range of software programs available for project management. One of the best, and the one used most widely by larger companies on large projects, is Primavera. Other currently available programs include Pertmaster, Open Plan, Microsoft Project and MacProject. They will even identify the person-hours

To make coffee with my cappuccino machine, my wife clears out the old coffee and loads the machine with fresh, inserts the cup into the machine, fills the machine with water, puts milk in the cup, switches it on, and when the coffee starts running through, froths the milk with the steam jet. She then sugars it, and stirs. It all takes 6½ minutes. I switch on first, quickly fill the machine with water before it overheats, and can get all the other things done before the coffee runs through. I also put sugar in with the milk to get it stirred with the frothing. Total time, 3.5 minutes.

I've had my leg pulled for being so fanatical about time, but our time on Earth is limited, and that's all we've got. If I make coffee twice a day, 360 days in the year, it saves 35 hours a year—a working week. Time saved in which I can entertain myself, or just be idle. If you want to waste time, best do it on purpose, not because you're being inefficient. So anything you have to do frequently, however simple, is worth planning properly.

needed for each activity, allowing the contract manager to plan his manpower resources.

Simple jobs don't need the full rigmarole of CPA. The most common and most efficient aid to planning them is the bar chart or *Gantt chart*. Figure 20.1 shows a simplified bar chart for building a house.

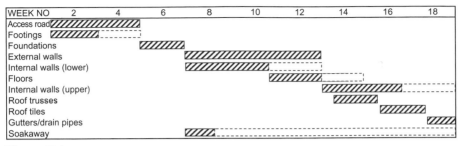

Figure 20.1   Project Gantt chart.

The top of Figure 20.1 shows the time during which the job is to be done. Having broken the project down into its component activities, each component is assigned a horizontal line on the chart. Bars on the chart then show the actual dates when the various jobs are to be done.

The first attempt at drawing this chart (Figure 20.1) might have "external walls" starting at the time that "foundations" finishes and "roof trusses" starting when "internal walls" are complete. It is obvious that these activities are in the controlling sequence. The gutters and drain pipes have to be installed after the roof tiling is complete, but the rainwater soakaway into which they discharge can be done any convenient time so is not part of the controlling sequence. But while the laying of bricks itself can't be speeded up, you could shorten the critical path by starting on bricklaying as soon as enough of the foundation is ready to make it worth while.

The chart shows this overlap with the dotted lines showing the *float* for the activity. Bar charts sometimes show two lines for each activity: one for "plan" and the other, which is filled in as the work proceeds, for "actual."

When we started a major extension to our small cottage, the weather was so awful that the work got badly behind schedule. The contractor (a qualified civil engineer, as it happened) came one day and said: "Your architect's plan has a reinforced concrete slab over the whole extension to form the first floor. We'd have to finish that before building the first floor walls and put the roof on after that. Shuttering, and reinforcement, pouring concrete, and waiting for that to set will take so long we wouldn't get the roof on before Christmas, when outdoor work has to stop until March. It means you'll move in three months late. I'll have a cash flow problem, and I won't have any work for half my staff for three months, so I'm likely to lose them. But I could use prestressed concrete planks instead of concrete. Your architect says he's happy with that. They cost more, but I'll bear the extra cost because it's worth it to me."

It went swimmingly. The planks turned up on a truck, together with a crane. They were lowered into position in a day. Laying bricks for the first-floor walls began at once, the roof was up and weatherproof well before Christmas, and we moved in at Easter.

I use house-building as an example because everyone can understand it and because I have an excellent example from personal experience. It demonstrates not only how to plan a job, but how—when things go wrong—it may be possible to get back on plan by spending a bit extra. This is called *crashing the plan*.

The great virtue of bar charts is that they are clear and easy to read at a glance. Software is now available for setting up a bar chart on the computer.

The contractor and the purchaser do not necessarily have the same goal when planning the job: the purchaser wants the project finished as quickly as possible, so that he gets the benefit from it soon, and only has to raise the money for the shortest possible time. It would be a good thing if the terms of payment in the contract were set so that the contractor has the same objective. In projects where the contractor only gets paid 90% of the contract price on delivery of the plant to the purchaser's site, he will be as keen as the purchaser to get on with it, and might even spend a bit more to accelerate the work.

There are (rare) occasions when a contractor might want to slow the job down. This could happen if a big payment is due from contract A, at a time when contract B needs a large sum of money spent on it. The contractor may deliberately slow down on contract B so that the money from contract A arrives before the bill for contract B does. Thus, the contractor avoids being caught with a cash flow problem between these two events. Some builders did this to me once, and I wasn't very pleased.

SUMMARY

- **The longer a project takes to finish the more money it costs.**
- **The terms of payment on a contract should be aimed at speeding up completion.**
- **Sometimes it's worth spending more to achieve this.**
- **Every complex set of activities has a critical path that controls the minimum time it takes: speeding things up means shortening that path.**
- **That could lead to another path becoming critical.**
- **There is a variety of software tools available for project planning and management.**

Chapter 21

Procurement and Monitoring

MORE PLANNING

Planning doesn't end with the creation of a pretty chart that shows what should be happening in the future. Planning is intimately connected with purchasing. The plan for the project gives the purchasing department the times at which the various materials and services will be needed. Ideally, everything should arrive exactly when it is wanted. Getting materials too late is obviously bad because it delays the whole job. Getting them too early is also bad because it may lead to problems of storage. If materials have to be stored, that in turn could mean double handling and yet more unnecessary cost. And, if goods arrive on-site early, so will the invoice for their cost, which is bad for cash flow. Services, of course, must be on time.

HOW (AT LEAST SOME) PROCUREMENT DEPARTMENTS WORK

Materials are purchased and subcontracts placed by the contractor's procurement department (or some similar name). This works like the purchaser in the main contract. First, the design engineers have to write a specification for the material they want to buy and then they raise a document called a *requisition*, which is an instruction to the procurement department to buy the specified material. The procurement department sends out tender documents (they usually send a copy of

> For a particular contract we wanted to purchase valves from a top manufacturer because we knew they would do the job, but our purchasing department insisted on competitive tendering and managed to save 10% by going to another supplier whose valves just met the specification. The purchasing department was given a pat on the back for its efforts. The site engineers, who subsequently had to replace all the valves because they failed, were given a severe reprimand for overspending.

Engineering Money: Financial Fundamentals for Engineers, by Richard Hill and George Solt
Copyright © 2010 John Wiley & Sons, Inc.

the main contract documents and expect subcontractors and even material suppliers to accept all the conditions of the contract). When the tenders are received, they are passed back to the engineer who wrote the requisition, and he prepares a tabulated evaluation (usually called a *bid tab*) to ensure that the comparison is on a "like for like" basis and decides which is the best option. There is sometimes disagreement between engineering and procurement because their objectives are different: The design engineer wants the best while the procurement officer wants the cheapest.

Having decided where to buy the material or service, they look at the contract program to decide when to place the order. They have to arrange with suppliers exactly when they are supposed to deliver. When the due date of critical items approaches, they will contact the supplier again—putting

> *"There is hardly anything in the world but some man cannot make a little worse and sell a little cheaper, and the people who consider price alone are this man's lawful prey."*
> —John Ruskin, English polymath

the "chasing" into "purchasing." And procurement departments should know which of their suppliers is reliable in this respect and which is not, so they know where they have to apply pressure.

All plans go wrong, some more, some less. So a vital function of the planning department is to monitor actual performance against the plan, and if there is a significant delay, get some action. Unfortunately, this is where many projects fail to get the service they need.

COMMUNICATIONS

The first problem is the speed with which information can be gathered and transmitted. The battle scene is the site where construction or fabrication is actually going on. The headquarters, where the planning department sits, may well be a long way away—either physically or functionally, or both. That can lead to long delays in transmitting information back to headquarters and orders to the front line again. Quick communication is essential. If the timetable looks as if it is going to get bogged down, then it's not much use finding out about it a couple of months later.

In the major battles of the Napoleonic wars, massed cannon firing with black powder generated great clouds of black smoke (gunpowder is expensive nowadays. People who reenact these battles today load their cannons with as much powder as will give a satisfactory bang and flash of flame, and they load some flour on top of that to yield the authentic amount of smoke). Once the battle was under way, the commanders couldn't see much of what was going on. By the time a dispatch rider had galloped to the scene and back again, it would probably be too late to give useful orders. Communication was so slow that some of the fighting was pretty well out of the commander-in-chief's control. No wonder Napoleon said he didn't want good generals—he wanted lucky generals.

Supposing, however, the news does get back quickly enough for useful action to be taken. The trouble is that "taking action" generally means spending money. The job was originally planned to incur the lowest cost, so any change from it will probably cost more. Action usually means bringing on extra resources, or a more expensive method of doing the job or putting staff on overtime, or something like that.

Maybe the person in charge of the job at the sharp end knows that at some modest cost he could overcome the delay and so save a much greater loss. But often he hasn't the authority (or the nerve) to spend the extra.

My company installed a boiler feedwater plant for a new power station. The power station operations staff had told the main contractor (our customer) the date at which they planned to fire up the boilers, and the main contractor gave us the date on which he wanted boiler feedwater. Unfortunately, he had overlooked the fact that cleaning the boilers also called for water purified to the full specification. We were already struggling to meet the date on our contract when we discovered that the water was really needed long before that. But no one suggested that they would make it worth our while to do something to reduce the delay. Maybe the project manager didn't have the authority to spend more.

THE S CURVE

The cumulative direct costs on a contract tend to follow an S-shaped curve (Figure 21.1)—starting at a fairly low rate, then increasing sharply as materials and subcontracts are procured, and then leveling off as the contract nears completion.

We've seen how the terms of payment define when cash will be coming in. Thus, we can plot a cumulative income curve for the contract.

Superimposing the cost S curve on the cumulative income curve shows where cash flow is negative (where the cumulative income line falls below the S curve) and where it is positive (where the cumulative income line is above the S curve), helping us to plan procurement so that we can ensure as positive a cash flow as possible.

Of course, real S curves don't look quite so perfect. Figure 21.2 shows the S curve for the $5.5 million contract that we looked at earlier (see Chapter 9).

The S curve shows only the direct costs on the contract and the cumulative income is based on stage payments as follows:

10%	With order
20%	On approval of drawings
50%	On delivery of all equipment
10%	On completion of installation
10%	On completion of commissioning

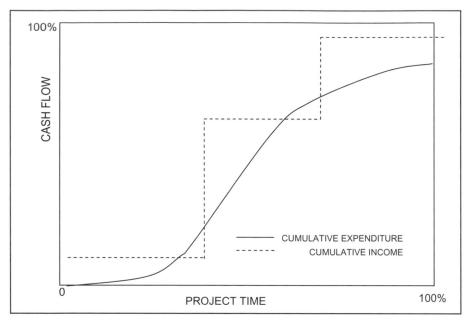

Figure 21.1 Contract S curve.

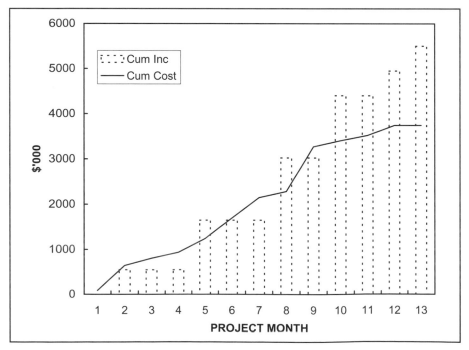

Figure 21.2 S curve for $5.5 million contract.

At the top of the S curve, the difference between cumulative income and cumulative cost is the contribution generated by the contract.

SUMMARY

- **Good planning consists of taking an overall view of the project and establishing the controlling path and/or doing a complete CPA exercise on it.**
- **If possible, set out the plan in some way that is easily understood—the bar chart is ideal for this if the project isn't too complex to show the plan clearly.**
- **Working closely together with the procurement department ensures that goods and services turn up at the right time and that money is spent in a way that allows control of cash flow.**
- **Monitoring progress as the work proceeds should provide a quick report of any delay.**
- **Have an organization in which the news of significant delay gets to someone who has the authority to take any necessary action.**

Chapter 22

Paying and Getting Paid

WHAT'S THE PROBLEM?

After years of painstaking research, I can reveal that no one likes paying bills. Also, if they do have to pay, they prefer to pay as late as possible.

This applies to all commercial transactions, but the problems are particularly acute in project engineering, where it is normal for part payment to be made in stages. The purchaser wants to pay as late as possible, and the contractor wants to be paid as early as possible. The battle is always on. Engineers may think that once they have completed the actual construction, getting the money into the bank is none of their business. In fact, they still have a vital role to play.

Getting paid is yet another procedure that can be divided into elements that have to follow one after the other, but it only has one path, every element of which is critical as Figure 22.1 shows:

1. The contract specifies when and how much the purchaser pays the contractor.
2. The contractor achieves the specified milestone.
3. The contractor issues an invoice for the amount due.
4. The contractor receives a check from the purchaser.
5. The cash arrives in the contractor's bank account.

THE CONTRACT

The terms of payment (which detail the amounts and stages at which payments should be made) are part of the contract and must therefore be settled before the contract is signed. Sometimes this arises during the final negotiations. I can remember one occasion when we had tendered for a large job and everything (including the price) was more or less agreed to and ready for signing when the purchaser's purchasing officer changed the terms of

Engineering Money: Financial Fundamentals for Engineers, by Richard Hill and George Solt
Copyright © 2010 John Wiley & Sons, Inc.

Figure 22.1   Critical path for getting paid.

payment in the contract documents—he reduced the down payment at the start of the project and increased the later payments to make up the same total sum. We had already been told that we were going to be awarded the contract, so this was psychologically hard to resist. It seemed too trivial a matter for us to start raising objections at this stage—it's not as if they were demanding a price cut!

But, actually, of course they were. If the purchaser were to pay a smaller first installment than the amount on which our price was calculated, he would reduce his finance costs and increase ours. It would also put an additional load on the contractor's working capital. We had a greedy sales executive at the time, and he accepted it.

ACHIEVING THE MILESTONE

When the contractor reaches one of the milestones set out in the terms of payment, he must demonstrate formally that the milestone has, indeed, been achieved. This formality takes various forms—a measurement by the quantity surveyor, or a certificate that the work has reached the agreed stage in the construction program, or (at completion of the project) passing some form of acceptance test. Whichever it is, some piece of paper must be signed by the project manager—who may, of course, be a consultant.

> A successful project depends on good cooperation between the purchaser's project manager and the contractor's contract manager. It is common experience that once a contract goes sour, it will go comprehensively sour because then relationships become strained even if they weren't strained before. Bad relations between the two can, of course, be the fault of either side, or of both, but the contract manager should have a better approach because it is an important part of his professional expertise.

It is essential for the contractor's team to realize the urgency of getting faults and omissions corrected so that the project manager has no reason for refusing to sign the certificate. If site relations are bad, for whatever reason, and the project manager feels hostile, he will insist that every last detail of the contract has been met, however unimportant it actually is. Things can get quite bitter.

The personal qualities of the contract manager are crucial. Suppose he has claimed that the work has reached a payment point, and he and the project manager are now going over the facts on-site. There will always be some minor faults and, however trivial or even imaginary they may be, they can be used as an excuse to hold things up.

When these are pointed out, if trust and confidence have been built up, the contract manager might say "OK, I'll get that fixed before the end of the week." The project manager will be confident that it will be done and may well agree that the payment point has been reached, even though he has nominal justification for refusing to sign the certificate.

Some purchasers devise complicated procedures before money is actually paid, but if such procedures have been agreed to as part of the contract, they have to be gone through.

> In the 1970s when the UK's electricity generator was state owned, it had a near monopoly of building power stations and could dictate the terms of contracts. One contract specified that the work had to be carried out "to the Engineer's satisfaction," and he insisted that the ¾-inch nuts on all flanged pressure vessels were squared up before he was "satisfied" and allowed us to be paid.
>
> On another contract they wouldn't let us test the chlorination equipment because the chlorinating plant hadn't been painted with the official color coding for water, chlorine, and so forth. But they wouldn't let us paint it because the chlorination equipment hadn't been tested. Oh yes, it was all in the contract documents.

REAL AND VIRTUAL MONEY

As far as the balance sheet is concerned, an invoice that has been issued is an asset to the company. In other words, as far as the accountant is concerned, an invoice is considered to be paid as soon as it is issued. Real life isn't like this. Most companies operate on a system known as *net cash monthly account*, which means that they will settle all invoices received before the end of a month at the end of the following month. Even this is often impracticable because, in large companies with computerized accounts departments, the payment procedure can take longer. So a contractor's invoice that has been issued is like virtual money. It's an asset but it's not actually there in the bank, so it can't be spent. It only becomes real money when the purchaser's check is in the contractor's bank.

When a company receives an invoice, it has to verify that it is correct. This means that the accounts department has to find out who in the procurement department issued the purchase order against which the invoice has been raised, and that person has to make sure that the engineer who raised the requisition is happy that the goods or services described on the purchase order

> The UK's state-owned coal industry, the National Coal Board, worked like this: The contractor's site engineer would get the acceptance paper from the NCB's project manager. He had to submit this to the local NCB area office for approval. They sent it back to the contractor's engineer on-site (second-class post, of course) who had to send it back to his head office, which issued an invoice that had to go to NCB headquarters in London. Maybe this rigmarole just served the NCB's bureaucratic nature, but it felt like just another excuse to delay payment.

have, in fact, been supplied. Often the accounts department is in a separate office or even a different town from the procurement department, and the engineer may be on the site, so this authorization cycle may take some time. In addition, many companies only issue checks at the end of a month (a computerized *check run*) so if payment of an invoice is not approved by, say the 27th of the month, the supplier will probably have to wait for approval until the end of the next month and get paid at the end of the month after that.

So far we've looked at payments from the purchaser to the contractor, but the contractor's cash flow also depends on his payments to suppliers and subcontractors, and so on down the supply chain. If a contractor is short of cash, he will often delay payment of invoices until the purchaser has paid so as to keep his cash flow positive. This means that the supplier, who has his own suppliers to pay, may have to borrow money to settle his own debts.

The difference between the real money (money in the company's bank account) and virtual money (debtors who appear on the balance sheet as an asset) is what cash flow and management accounting are largely about.

Most contracts will set out the payment period—the time between receipt of invoice and release of payment. Usually this will be 45 days, although sometimes it's 30 days, 60 days, or even 90 days. Obviously, the payment period is critical to the contractor who may have to borrow from the bank to cover the virtual money. A smart contractor will agree to longer payment periods with his suppliers and subcontractors than those in his contract in the hope that the virtual money will become real money before they have to be paid.

Back in the 1960s Arnold Weinstock, CEO of GCE, the UK's biggest electrical engineering company, discovered that he could earn a lot of interest by getting his customers to pay a few days earlier while he delayed payment to his suppliers by a few days. Getting £500,000 from a customer a week early while making a supplier wait a week for £500,000 means that the company has an extra £1,000,000 on deposit at, say 5% per annum, for a week. That is almost £1000 for doing nothing.

Other companies adopted this system. It became common practice and the few days became weeks. The result is that many large companies made a lot of money at the expense of small suppliers who had to borrow money to pay their suppliers. The added debt burden drove a lot of them out of business, even though their balance sheets looked good.

BAD PAYERS

This is the theory. In practice, late payment is all-too-common and is often passed on all the way down the supply chain. The effect can be disastrous, especially on the little fellows at the bottom (see Chapter 9). They haven't got large capital reserves, can't raise a loan from their bank, and they go bust.

Several nations have laws intended to discourage late payment by entitling the creditor to claim quite punitive interest rates for a period during which the payment is legally overdue. I am not aware that it has made much difference in practice. Suing a purchaser generally means fighting a bigger and richer company, and only the lawyers are likely to profit from it. It is also unlikely to encourage the purchaser to employ the contractor in the future.

A system of payment mostly used in large international contracts is worth mentioning. At the time that the contract is negotiated, contractor and purchaser enter into a three-party agreement with a bank, and the bank makes the money available as it is needed. Then the money goes from the bank directly to the contractor, but the five stages still have to be gone through.

A final note: Sending checks means delays: in the post, getting the check to the bank, and before the bank actually credits it to the account. When I was a company director, I would often go by the bank on my way home to put the check into the night safe deposit and so save a day's delay. Electronic transfer between accounts should be quicker, but with many banks it takes 3 days. Where is the money during that time? In some banking limbo between purchaser and contractor? Considering how much money goes through a bank every day, it must be a rich limbo, capable of earning a lot of interest for the enforced delay. No one has yet explained to me why this delay is necessary, or who gets the benefit.

Large companies quite often used their muscle to retain money in their account when it was due to a smaller company that lacked the clout to demand payment. I worked as a design consultant to a large civil contracting company. Although net cash monthly account was agreed to, the contractor did not pay on time. When an invoice had remained unpaid for 3 months I went to see the contract manager. He apologized and said that he had authorized payment of the invoice 2 months ago. Together we went to see the accountant. He asked for details of the invoice, checked his records, and opened the top drawer of his desk. I was rather surprised to see it was full of checks. He flicked through them quickly, pulled one out, and handed it to me. It was for the full amount of our invoice, signed ... and dated 2 months earlier.

I went to a wine auction to buy some cases of claret that I was to split with the co-author of this book. After the auction I rang him and told him how much he owed me. "I'm sending you the check today," he replied. Next morning, no check in the post—why? Because, like any good businessman, he had put a second-class stamp on it. We authors know how to put theory into practice!

In medieval times you harvested your corn, took it to the miller for grinding, and he would return it to you as flour. It was widely believed that millers were rascals who got rich by keeping some of it, but there was no way of proving that. So, although their services were indispensable, millers were generally held in low esteem. Just like bankers today.

As you'll have noticed, I have used examples of bad payment from the 1960s and 1970s, when for a time it became common UK practice and caused great hardship. When my Finnish colleagues realized that, they simply put a surplus on all their quoted prices to British customers. I think we've learnd from this that bad payment is bad for everyone.

SUMMARY

- **Everyone wants to pay late and get paid early.**
- **Contracts must set down exactly who pays what and when.**
- **That still leaves many ways of delaying payments legally or even illegally.**
- **Late payments by big purchasers can cause problems all the way down the supply chain.**

Chapter 23

Consultants

BRIEF HISTORY OF ENGINEERING

We've had a brief history of money. In a book about engineering and money, it's only fair to give engineering the same treatment.

The ancient Egyptians must have been tremendous engineers, but we don't know much about how they did it. The Romans achieved great things with aqueducts, heating systems, and so on, but their engineering aimed mainly at military ends: building roads to carry armies and their *engines of war* (catapults, rams, and so on).

These military objectives ruled until the eighteenth century, when General Wade built great military roads across Scotland, which were intended to subdue the Highlanders (I don't think anyone has really ever managed that, to this day).

> I pointed out earlier that the word *engineer* comes from Latin *ingenium*, which also gave us the word *ingenious*, and that reflects the fact that the essence of good engineering is creativity. Sadly, the word *engine* has in English come to mean a machine or motor, which has dragged the word *engineer* down to include mechanic, machinist, fitter, and train driver.

> The tradition of military engineering had never really died. The Corps of Royal Engineers claim that their history goes back 900 years. The Royal Military Academy at Woolwich was founded in 1741.

Late in the seventeenth century, however, the Industrial Revolution began in England with the building of canals. They meant that for the first time in history we could carry bulk materials across land, and so bring iron ore and coal together in the same place. The first canal builders were brought over from Holland, but in England they faced an unfamiliar landscape of rolling hills that needed novel engineering solutions—locks, lifts, and transporters. In 1790 John Smeaton, the "father of canal building" founded the Society of Civil Engineers, choosing that name because it was intended to embrace all nonmilitary engineering.

Engineering Money: Financial Fundamentals for Engineers, by Richard Hill and George Solt
Copyright © 2010 John Wiley & Sons, Inc.

The canals, however, had led to a huge upturn in the production of iron. The great ironworks of Merthyr Tydfil in South Wales made more iron, and cast more cannon, than the rest of the world together—a great help in beating Napoleon at Waterloo.

With coal and iron so freely available, Newcomen, Watt, and Trevithick developed the steam engine as a source of power, first for pumping and later for operating machinery in spinning and weaving mills. Stephenson did the same for railways, and the railway age produced the Brunels. It took a century for the world to catch up with, and eventually overtake, England as the engineering center of the world.

The organization that was to represent nonmilitary engineering looked good—except for one problem. The civil engineers thought of themselves as gentlemen, far above the coal-blackened, oily-fingered builders of steam engines. They refused to accept them into their society, which led the men with black fingernails to form the Institution of Mechanical Engineers, which eventually restricted "civil engineering" to its modern meaning.

Most continental European nations have a single engineering body to confer professional status. There qualified engineers are professionals on a par with doctors and lawyers. Not in Britain. Once we had the two engineering institutions, other branches such as the electrical and chemical engineers set up their own groups, along with a host of other specialist institutions, some with a doubtful claim to professional legitimacy. This fragmentation meant that engineering was

A 1775 map of London shows two steam-powered water pumping stations, one described as "An Engine for raising Water by the Power of Fire." There was also an amazing pumping station on a floating base fixed to London Bridge. It had to rise and fall with the tide, and it was powered by a water wheel that used the fall created by water passing through its narrow arches. Taking a boat under the bridge was like shooting the rapids, and people kept getting drowned.

This is the London Bridge that was "falling down" in the nursery rhyme, and it really was. As its piers kept slipping, they would reinforce them from the outside, which narrowed the width available for flow, which increased the scouring action of the water, which made the piers slip some more. Then they decided to lighten the bridge and restore the original width for the flow, only to realize that if they did that the city would have no water supply. People just had to keep getting drowned until a new supply from Sadlers Wells could be commissioned early in the nineteenth century.

Here's how rivalry and snobbery operated to fragment the profession: I graduated in 1950 and was looking for a job. They told me not to bother with ICI, then by far the biggest chemical company in Britain. ICI did not recognize chemical engineering as a profession because the Institution of Chemical Engineers (founded in 1922) was considered a Johnny-come-lately by the "senior" institutions. Chemical engineers are now recognized as a superior species—salary reviews show that we earn more than the other sorts!

never defined as a profession, and in the UK the description includes the man who is sent to mend your washing machine.

The situation in the United States is not quite so bad. However, even there the driver of a railway train is an engineer.

The most celebrated Victorian engineers were the Brunels, father and son, whose pioneering ventures include a tunnel under the Thames, the Great Western Railway, and the huge steamships *Great Western*, *Great Britain*, and *Great Eastern*. It is astonishing that they could do all that alone. By contrast, major ventures since then need a variety of specialized disciplines to work together. Engineering consultants began to appear.

> I committed a minor traffic offense in Italy and was stopped by two *carabinieri*, who demanded to see my passport. This was the old sort that showed your profession, and it said "engineer." They nodded appreciatively, "Ah! Inginiere! Bravo, bravo!" It would not have happened in England.
>
> Then my wife stuck her head out of the car window and said—show them your date of birth! It happened to be my birthday, and I showed them. At this, they laughed, and asked me round to a nearby *trattoria* for a drink. As I say, it would not have happened in England.

WHAT'S A CONSULTANT?

Until the Brunels, the most successful engineers were also entrepreneurs, and some became quite rich. But as the projects became bigger and more sophisticated, they were financed more and more by corporations, so these opportunities dwindled. Engineers who couldn't start their own businesses had to turn to selling their skills by becoming employees—or consultants.

Contractors and consultants have rather similar functions: They are paid by a client to take a part in executing a project. In fact, their skills overlap at many points. Neither of them needs much capital to set up in business because they are both selling skills rather than material things. A contractor supplies material, true, but only things that he has bought or had fabricated. He also supplies labor, but again much of it is hired for a specific project. Neither kind of activity needs much fixed capital: In fact, consultants, who sell only their skills, need even less working capital than contractors.

> In London it is common for trades to collect in the same area or street, such as doctors in Harley St and jewellers in Hatton Garden. In the nineteenth century engineering consultants settled around Victoria Street, Westminster, like the young engineer in the Sherlock Holmes mystery *The Adventure of the Engineer's Thumb*. (As I recall, he wasn't very successful.) There are now 3000 doctors in Harley St, and jewellers still thrive in Hatton Garden, but I can't think of any engineering consultants still in Victoria. Maybe they can't afford the rent.

Big industrial companies used to carry staff who would deal with general engineering problems. More recently, companies have shed employees who are not expert in the company's own technology. So now, when they consider any project, they need outside help, which should mean more specialist consultancy work. Indeed, even big consultants themselves often employ outside specialists in areas that are not in their field of expertise. Sadly, there are all-too-many examples of projects that didn't get such help because the client was too stupid or too stingy to pay for it, and found that that was a false economy (see Chapter 12).

The name "consultant" has come to represent two rather different kinds of enterprise:

- Big ones who offer a range of technical and organizational skills
- Small ones, often only one-man affairs, usually with some expertise in a specialized branch

BIG CONSULTANTS

Looking at the list of stages through which a project goes, you will find that the big consultants can do everything needed from the point at which the client decides to commission the project. The consultants' duties on it can include some or all of the following:

- Preparation of tender documents
- Tender evaluation
- Recommendation for contract award
- Site supervision
- Remeasurement (in civil work) and valuation
- Agreement of extras
- Release of payment to the contractor
- Witness testing of equipment
- Acceptance of completed works

Typical fees for this kind of work would be 10% of contract value for smaller contracts, falling to 5% for big ones. Or they might charge by the person-hour, at £50–£200/hour (in 2010) depending on the grade—not astronomical rates when compared with other professions. Consultants working on a percentage of the contract value have a vested interest in seeing the highest capital cost. They will often favor overengineered designs with higher contract prices.

Originally, the big consultants were civil engineers. Now they undertake all kinds of work, though something of the civil engineering bias remains even with those who can undertake a sea-water desalination plant as easily as a complete airport.

If consultants are commissioned by the client to do everything, the contractor is reduced to an artisan who just does as he is told. Since no skill is required of him, he will generate none, and the technology remains static. At one time, U.S. consultants in my field of technology were so dictatorial in their specifications that no novelty was allowed to creep in. Result? European (particularly German) practice was then 10 years ahead.

In the big consultants' heyday the consultant *knew*, whereas the contractor didn't, and it bred a certain arrogance in them. Today the contractors often have in-house experts every bit as good as the consultants, and in specialized work they are inevitably more up to date. But, in the UK at least, the arrogance hasn't yet quite disappeared.

All this bred a confrontational attitude: some consultants still see contractors as unscrupulous tradesmen who swell their profits by cheating whenever they can, and some contractors see consultants as arrogant spectators who thwart their efforts to make an honest living. Usually, but not always, they're both wrong.

Traditionally, where a big consulting firm was managing a project, they appointed "the Engineer" (usually a partner) to be in charge, with awesome power to judge whether to accept work and (more importantly) whether to release a payment for it.

> Where the consultant writes the contract, he can also write the rules of the game. Very sensibly, he will try to protect himself by passing responsibility for technical problems on to the contractor. The contract might read "notwithstanding the foregoing the contractor shall be responsible for providing a complete and fully operational plant," that is, in plain English, "if we're wrong, it's the contractor's fault." Elsewhere it might say "the contractor's scope of supply shall include but shall not be limited to …"—again, in plain English "if we've forgotten anything, the contractor must pay for it." After years of lawsuits, these tricks no longer offer the consultant much protection, so now he must shoulder his responsibilities or officially pass them on to the contractor.

> The clear implication is that "the engineer" is the only real engineer around, and he assumes a God-like stature. That mindset is (hopefully) dying out.

SMALL CONSULTANTS

To divide consultants into big and small is, of course, artificial—there is a size continuum. But there is a very large number of really small ones, many of them one-person outfits. They trade on being specialists in some field, and some may also trade on having the specialist equipment that is essential for it. The technical variety of subjects that all these people cover is huge, ranging from soil mechanics and geophysics to biotechnology and precision machining.

It is not a route to easy riches. A small consultant sells intellectual property, measured in terms of person-hours. No one can actually book more than 1500 hours/year to his contracts (see Chapter 13). Out of his gross income he has to pay his overhead, which includes an office and all that goes with it, a car, and maybe some specialist equipment. He has large telephone bills; he needs to pay for help with accountancy and tax. Professional indemnity insurance is mandatory and can cost anything from £1000 to £10,000 a year. Vacations take up time when he might be working and earning money. When he spends a morning at the dentist he can add the loss of income to the dentist's bill—always provided he gets enough work to fill his time but, like a contractor, he'll be lucky if he has much in the way of steady work. Enquiries come in at random, with periods of overwork alternating with idle ones. He's also one of a group of people who are hit first and hardest when there is an economic downturn.

> I used to wonder why engineering consultants have to charge much lower hourly rates than, for example, management consultants. The head of one of our most prestigious schools of management (who calls us engineers "plumbers") explained it to me. "The less precise your subject is, the more you can charge" he said. "Your trouble is that you deal with facts and numbers. Nobody pays a lot for that. Management is always in a bit of a fog, so management consultants can charge more. But the guys who give seminars on really woolly stuff like 'leadership' or 'lateral thinking,' their rates are astronomical— and people pay them. Fantasy is worth more than fact. And that's a fact!"

This end of the consultants' range has a particular problem. A consultant trades on his expertise, which means he must have experience. Who is better equipped for that role than some highly experienced and recently retired technologist? He may have retired (prematurely, perhaps) with a good pension. He doesn't actually need to earn a living any longer, but thinks it would be nice to make a few cents more. So he becomes a consultant (which is what my co-author has done). There are many like him, so it's a very competitive market, filled by people who don't really need to charge a lot. It keeps the hourly rates low (and, as an independent consultant below retirement age, I suffer as a result!).

CONTRACTORS

Another problem is that specialist contractors have the same skills as consultants and may even be more up to date. In order to get contracts, some of them give advice free. Of course, this advice is bound to be biased—at worst it could be like getting a fox to tell you how to build a chicken coop. But the client should take this free advice seriously—at least it's likely to be up to date —while remembering that there is no such thing as a free lunch. The advice is being given to them because the contractor hopes to get the job. That's OK, but is it really sound, or is the contractor just promoting something that only his company can provide?

Contractors themselves will sometimes take on consultancy. My old company sold some contracts to South Africa on a "design and commission" basis. On those contracts the client bought and constructed everything to our specifications and designs. When they'd done that, we sent out our commissioning engineers to get it going.

In fact the lines between clients, contractors, and consultants are becoming more and more blurred. The newer industries such as computer software have led in this trend. There are now huge international contracting companies that have all the necessary in-house expertise for a wide variety of projects. (When they haven't got it all, they take on specialist consultants to fill the need.) Some of them (such as Uhde or Kellogg, Brown & Root) are true multinationals with offices in major cities around the world. The purchaser can award these contractors a "turnkey contract," according to which the contractor does absolutely everything.

The relationship between large contractors and large consulting firms takes various forms. Sometimes the specification and design work remains to be done by a consultant. Sometimes it doesn't, even though the contractor may have all the skills needed to do it himself. Sometimes the consultant's role is limited to checking that the contractor is doing a good job.

In cases where a joint venture consortium has been formed to finance a project, the banks who back the project may insist on a consultant to supervise the work, for fear of getting stitched up by these naughty technologists. In one of my court cases (see Chapter 12) the consultants had been taken on because the bank demanded it, not because the contractor or purchaser wanted them. Their real duties were hardly more than nominal.

Moral: Consultants provide an essential service. They, too, think they are entitled to make an honest living, and they will do only that work for which they are paid. Like the contractors who will build a bad plant if the specification demands it, a consultant will do a poor job when he's not paid properly.

"Pay peanuts, get monkeys," they say.

The Russian version of this in the days of the Soviet Union was "They pretend to pay us, and we pretend to work." Unsurprisingly, people are the same everywhere.

SPECIFICATION

The most important part of any project is the preparation of a specification, and this is often the consultant's duty. When the specification is wrong, a good contractor will try to persuade his client of that, but in the end he must carry out the work specified in the contract—as long as he has complied with the specification, he is in the clear and maybe he can do the remedial work as an extra to the contract. Employing a competent consultant—either an individual or a firm—to ensure that the specification is correct before placing the contract is generally cheaper than making a mistake. And the hourly rates charged by consultants for engineering work are a lot lower than the rates charged by the same consultants for expert witness work in litigation.

Many purchasers will accept the "free" advice given by contractors rather than spend money on employing a consultant. Often this advice is perfectly sound, although it may be loaded in favor of some process or construction technology unique to that contractor.

Some contractors, particularly in the process industries, are offering their specialist services as consultants, while many traditional consultants are either forming strategic alliances with contractors or are themselves taking on the role of management contractors. This illustrates how the roles of the contractor and consultant are becoming more and more similar—just as they have always been in industries such as computer software writing.

SUMMARY

- **Most sizable projects today call for such a variety of skills that specialist are needed.**
- **Failure to take on expert advice can be costly.**
- **Consultants provide unbiased advice. Specialists working for a contractor will inevitably favor what their own company can provide.**
- **The boundaries between consultants and contractors are becoming more and more blurred.**

Chapter 24

Using Your Judgment

CHOICES

Technology is about measuring things and working out the best solution on the basis of that information. It would be a relatively simple affair if the information were complete and exact, but it never is. Quite a lot of engineering is about coping with the problems that that shortcoming poses. It means using your judgment.

For the engineer, designing anything implies making different kinds of choices. The most basic choice is between completely different ways of solving some problem. For example, a road crossing of an estuary can be achieved with a suspension bridge or a box girder bridge. But, equally, it is possible to route the road further inland and so avoid the need for a large bridge altogether.

We are often faced with the kind of decision that must be made when there is a range of sizes available, such as, for example, selecting the optimum diameter for a pipeline or for a power transmission cable. Yet another arises when the problem is to satisfy a growing demand, and the engineer must decide for how long in the future he will cater.

LIFE-CYCLE COST

We defined money as a measure of all valuable resources. If that is the case, good engineering should provide the cheapest answer to whatever problem is to be solved. The cost we should aim to minimize is the life-cycle cost—that is, not only the cost of providing the item we are designing but also the cost of operating and maintaining it in the future. As I said above, the main problems here are: first, the numerical information is never really reliable, and, second, there are important influences that don't come in the form of numbers.

Engineering Money: Financial Fundamentals for Engineers, by Richard Hill and George Solt
Copyright © 2010 John Wiley & Sons, Inc.

Dealing with the numerical information first: The first cost of any piece of hardware can be accurately known—if you already have an offer from a contractor or supplier, you will know it with total certainty. But, the future costs of its operation and maintenance cannot be estimated with any confidence—power costs and labor costs will change with the years, just to mention two important ones. In many building and civil engineering examples the future costs should be relatively small, but in all other branches the future costs tend to far outweigh the capital expense.

ENVIRONMENTAL COSTS

We are at long last taking the environment seriously (see Chapter 26), so in making our choices, we must include the environmental costs of manufacturing, operation, cleaning up discharges, waste disposal, and final decommissioning and disposal of the equipment at the end of its life (particularly serious for nuclear plant).

Environmental costs are particularly awkward because they depend on the boundary within which we are doing the calculations. For example, we can consider the factory wall as a boundary and consider only the environmental costs associated with operation within it. That does not include the environmental costs of transporting chemicals and other consumables to the factory. If we draw our boundary say 50 miles around the factory, we might pick up the fact that the trucks supplying the consumables will cause problems in the local village, requiring road widening, pedestrian crossings, and so on. And even if you can define just where the boundaries are, most environmental costs can't be reduced to numerical data. In time, perhaps, we shall develop carbon footprint cost rules that are practicable for that kind of problem, but they are only part of the story.

Electricity is the cleanest form of energy—true or false? If Los Angeles were to allow only automobiles powered by rechargeable electric batteries, would that be an environmental benefit? That depends where: Los Angeles would certainly be less polluted but, all other things being equal, the total power demand could be greater. In addition to the power to drive the automobiles, there are losses in rectifying and battery charging, and in the transmission lines from a distant power station. If that is oil fired, then globally the CO_2 emissions will be higher.

Moreover, the electric automobile, with its battery and charger is (at present) more expensive than the gasoline-powered alternative. That implies that more materials and power have gone into making it, so it has a bigger environmental impact even before it leaves the showroom.

In my student days, we were given an exercise to calculate the optimum lagging thickness for a steam pipe. The variables included steam temperature, lagging conductivity, pipe diameter, costs of lagging per unit volume, and cost of heat. We combined these into an expression that calculated the operating cost for 20 years and equated the differential to zero to get the optimum lagging thickness. No one told us that the result was wildly irrelevant, not least because lagging comes only in standard thicknesses.

So there will always be some important parameters that can't be known with any confidence and others that can't even be expressed in terms of numbers—environmental problems are particularly awkward in that respect. That shouldn't stop us trying to calculate and optimize costs, but we must bear in mind how reality can confuse the issue.

OPTIMIZATION

In some engineering problems there is a continuously varying factor—such as, for example, when the designer has to choose the best diameter for a power transmission cable. It is possible (at least in theory) to take the factors that influence the life-cycle cost and gather them into a mathematical formula, differentiate it with respect to the cable diameter, and equate it zero to find the best solution. We now have software packages that do all this for us. However, all too often, they calculate costs over a range of alternatives and display only the option that yields the lowest cost.

You can have little confidence in any such exercise. Differentiating and equating to zero defines the "optimum" as the point on the cost curve where the slope is zero—that is, where it's flat. That pattern is common to all optimizations, including those that can't be reduced to a formula. And if you look at the cost curve, you almost always find that on one or both sides of the optimum point the curve stays flat for a wide range, so that the costs vary little even if you move away from the optimum.

But the assumptions on which these cost calculations are based can never be accurate. To be realistic, therefore, the cost curve should not

Here's a made-up example to show the point. You've chosen what small automobile to buy and must decide whether to have a gasoline engine or pay $1500 more for the diesel version. Both fuels cost $1 per liter, and the diesel gives 20 km/L compared with the gasoline engine's 15 km/L. You drive about 30,000 km/year, so the diesel will cost $500/year less to run, and you'd get your money back in 3 years—a decent investment. But you hope in future to work more from home, which will reduce the annual mileage. Decisions! Decisions!

About 10 years ago I was shown a major study into the design of a huge seawater distillation facility for Singapore, an island that is understandably nervous at depending for its drinking water on a pipeline from mainland Malaysia. Sure enough, the study selected an optimum design but sensibly printed out the costs for the entire range of options with respect to a main parameter (I forget which). Scanning the list, it showed very little difference between different designs—the difference between the highest and lowest cost was less than 2%. Fuel costs are, of course, the biggest single item of cost, of distillation, and the oil price has more than doubled since then. The identical calculation based on today's oil price would probably point to a completely different design as the optimum.

be a line at all but a cloud that takes into account the uncertainty of the data used. And that's even before we consider all those factors that cannot be measured.

COMMON ENGINEERING EXAMPLE

Let's suppose we want to pump 2.6 million U.S. gallons per day ($417 m^3/hr$) of drinking water along a 1¼-mile (2-km) main. A small-diameter pipe will be cheaper to install than a large-diameter pipe, but the velocity and, hence, the head loss due to friction will be higher in a small-diameter pipe than in a large one. As the pipe size increases, the capital cost increases and the costs of pumping decreases. So what is the optimum pipe size?

Lets assume that the capital cost of the pipeline is to be amortized over 25 years, that power costs 5¢/kWh, and that the pipeline will be utilized for 70% of the time. We can set out the costs in a spreadsheet as shown in Table 24.1.

It's fairly obvious from Table 24.1 that the 250-mm pipe yields the lowest cost. We can show this graphically (Figure 24.1).

This looks convincing, but you will note that the curve is fairly flat from 200 to 300 mm. You also have to remember that the capital cost (capex) curve

Table 24.1   Capital Cost of Pipeline

Pipe Diameter	(mm)	150	200	250	300	350	400
Pipe cost per meter	($/m)	120	200	280	370	470	580
Installation per meter	($/m)	240	400	560	740	940	1160
Total per meter	($/m)	360	600	840	1110	1410	1740
Capital cost for 2 km	($ 000)	720	1200	1680	2220	2820	3480
Amortized over 25 years	($ 000 pa)	29	48	67	89	113	139
Velocity	(m/s)	6.5	3.7	2.4	1.6	1.2	0.9
Total head loss for 2 km	(m)	463.3	103.3	32.3	12.5	5.6	2.8
Pump power at 70% efficiency	(kW)	751	168	52	20	9	5
Annual consumption	(MWh)	4608	1028	321	124	56	28
Annual operating cost at 5 ¢/kWh	($pa)	230	50	20	10	3	0
Total annual cost	($ 000 pa)	259	98	87	99	116	139

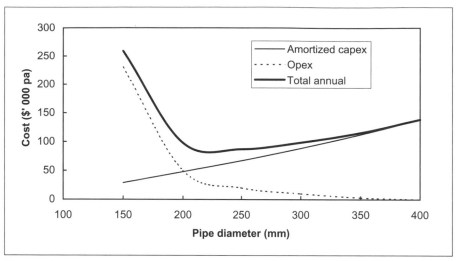

Figure 24.1   Cost of optimization curve.

is accurate because we have firm prices from contractors. The operating cost (opex) curve, on the other hand, is only accurate for the present—we don't know how the cost of power will change over future years. Another consideration is that the capital cost usually comes from one budget, while the future operating costs will come from future revenue. Given all that uncertainty, the project manager may well opt for the lowest capital cost he can get away with so that his project management looks good—in other words he'll probably buy 200-mm pipe. Or, if he's a municipality engineer with no budget constraints and who expects a future increase in demand for water, he might choose the 300-mm pipe. So optimum depends on a lot of things besides the simply technical.

CURSE OF TOM THE COMPUTER

The invention of the silicon chip has put a huge amount of computing power on the desktop of every engineer, but there is a danger that the availability of all this information may actually distract us from using our judgment. Computers are wonderfully quick and accurate calculating machines. Unless you

I went to my first computer course in 1957—early days! The computer occupied the whole top floor of Imperial Chemical House in Millbank (the top floor, in order to facilitate cooling, because it was all transistors). But the people who ran it had divined its character. "We call the computer TOM" tell it to do otherwise, the software will normally give you results to seven significant figures, which is nice of it.

Less nice is the fact that idle and foolish people (who are, sadly, the large majority) use that seven-digit number without questioning what it really means. They should look at the dodgy basis of their calculation, estimate its intrinsic errors, and present the result accordingly.

When designing a new model of automobile, for example, you might calculate its weight by adding all of its proposed components. Suppose the computer comes up with a total of 1351.530 kg. What does that really mean? A figure of 1350 kg might be reasonable, implying anything between 1345 and 1355 kg—three significant figures, not seven. But maybe the components' weights themselves are also estimates. Thus, it might make more sense to come down to two significant figures, and report that the automobile, if built, will weigh between 1300 and 1400 kg.

This is not a trivial point. One of the court cases in which I acted as expert witness (see Chapter 12) arose from someone assuming that he had been given exact information, when he hadn't. The outcome was that his company went bust.

IS IT MEASURABLE?

As I said earlier, all data are more or less inaccurate. Of course, this applies to factors in design that can be presented in terms of numbers, but—worse—there are also factors that are important in design that are hard or impossible to measure in

they said, "That stands for Thoroughly Obedient Moron."

The only change in 50-odd years is that today they're not quite as obedient as they used to be.

In the days before personal computers, engineering calculations would largely be done on the slide rule—affectionately known as the Guessing Stick. This had two advantages—the result couldn't be read to more than three significant figures, and it didn't give you the decimal point. You had to follow the calculation in your head and put that in yourself. Which, in turn, meant you had to follow the calculation instead of letting the machine do it. Also, if you had made any mistake in feeding in the information, you would probably spot it quite soon.

"Excessive accuracy shows mathematical immaturity"
—(*J.K.F. Gauss, German mathematician, 1777–1855*).

Few engineers are ever taught that we use numbers for three quite different purposes:

- *Numbers* can be used as names, for example, a Boeing 747, or the number 47 bus.
- *Integers* are for counting, for example, the number of people in a room. They are exact, whole numbers.

terms of numbers at all. As they just make life more difficult, there is therefore a danger that they get ignored.

UTILIZATION

One of the most important factors in estimating the life-cycle cost of any item is the proportion of real time during which it is actually going to be in use. This is an absolutely crucial factor that is sometimes ignored completely. It is a useful discipline to make it a rule to calculate operating costs in terms of dollars/year. You can't do that without supplying the calculation with a number for utilization.

When equipment is in use, it usually runs up operating costs—by using power, chemicals, and the like. Maintenance and supervision costs may also be linked to the actual time it is in use, such as, for example, the need for a 10,000-mile service on an automobile. When not in use, on the other hand, it may be costing nothing except the depreciation on its capital value.

Even when utilization is taken into account, the estimate is often wrong, and admittedly it is often difficult to predict. The utilization of equipment in industry is often unexpectedly low. One would think that the major units in an oil refinery, which works the year round, would be something like 50 weeks a year. Actually, records show that oil refinery units typically operate for only three quarters of the time.

In many cases it's not so much difficult as impossible to predict the

• *Real numbers* are measurements that are never completely accurate—people have calculated π to billions of places, and it still goes on. The important thing is to remember that real numbers are never totally exact. There is always some error, however small.

When designing an industrial water treatment plant, we were often asked to include in our offer the operating costs due to the consumption of chemicals in terms of cents per ton of product. That yields a number in which the utilization plays no part and that is not much use beyond comparing different contractors' offers for the job.

The cost in terms of dollars/year would be better. If nothing else, that would force the purchaser to think about the probable utilization and specify it to give the contractors a base for their sums. The result would show the purchaser the relationship between capital and operating costs. He might then find that the operating cost is relatively small and that a design low in capital but high in operating cost would be better. Or the reverse might be the case. Cents per cubic meter doesn't give you that information.

An automobile's job is to get you from point A to B, and the average family car does about 10,000 miles per year. Suppose its average speed overall is as low as 35 mph, that means 286 hours/year or 3.3% of the time—the other 97% of the year it just sits around rusting quietly away. This doesn't, of course, apply to the CEO's Cadillac. Its job includes sitting gleaming in the reserved parking space to prove to the world what an important and powerful man its owner is, so its utilization is much higher.

utilization. A manufacturing unit might be designed on the basis that it will operate during normal working hours. If it turns out that its product is successful and in great demand, will it then go to weekend working, or two shifts a day, maybe even three shifts a day? How is one to predict its utilization?

Utilization has a vital effect on the life-cycle cost figures—the higher the utilization, the greater the impact of operating costs in relation to the capital cost. With something that is only in occasional use, on the other hand, the operating cost is much less important. An extreme example: There is no point in buying an expensive efficient motor to drive a pump for a sprinkler fire prevention system. With luck it will never be used except for testing the system, so the annual electricity consumption is just about nil— a characteristic of anything that serves as "insurance" or "emergency" equipment.

Utilization has a direct effect on the engineering design of any piece of equipment. My automobile gave out after 7 years, at 126,000 miles. I thought that wasn't bad, but a long-distance truck might do the same distance in 6 months, so the specification to which the truck's engine is built is very different. (I'm told they have chromium-plated cylinders.)

Many years ago the government of Jersey in the UK's Channel Islands commissioned the building of a seawater desalination plant. The aim was to provide an emergency water supply to save the tomato crop in the event of drought. For 20 years there was no drought. By the end of that time, the plant had corroded into dereliction. It had never operated at all.

Usually, utilization refers to time, but here's another way of increasing utilization in order to get more work out of a piece of equipment. Many breweries nowadays brew the beer at up to double the strength at which it is to be casked or bottled, and water it down as the last stage in production.

I once asked a manufacturer of galleys for passenger aircraft what criteria of reliability was used, thinking of the disaster if the crew found in midair that they couldn't heat up 200 frozen airline platters. He said the aircraft makers' spec asked for a minimum period of (I think) 100,000 flying hours between failures, which is probably longer than the lifetime of the plane. So they do have a number to work to, but in practice it is little help: 100,000 hours is over 11 years—you can't realistically test cookers for 11 years on the off chance that one might fail.

RELIABILITY

Automobile engines have coil ignition. Light aircraft engines traditionally have two independent magneto ignition systems with two spark plugs per cylinder. Magneto ignition will function even if the battery is flat, whereas coil ignition won't. In case something does go wrong, a twin system means the chances of the ignition failing are very small indeed. The reason for spending so much more is obvious: Failure of an automobile engine is inconvenient, that of an aircraft engine is rather more serious.

Unfortunately, it is more or less impossible to express the need for reliability in numerical terms. And the level of reliability needed can vary a great deal. Any machine will fail sooner or later—it's just a case of when. A bridge should not fail, but sometimes it does. In any project, the specification should give at least a clue as to how important reliability is going to be. That leaves the problem of how to tie the contractor to some minimum standard.

It's not only total stoppage that comes into consideration. It could be that the service that the project gives might be allowed to suffer deterioration (of output or of product quality) for a limited time—sometimes this time is specified. For example, it might be the length of time that it would take to replace the cause of the fault. Or it could be that quality and quantity of the output must be maintained at all times.

> My wife had trouble with her teeth, and a series of dentists put in a series of bridges. We called these Tay, Tacoma, and Yarra, after those famous bridges. That showed some foresight, because in the end they all broke.

The responsibility for this is squarely with the purchaser or his consultant. He must either give a clear idea of the reliability required or specify the detail that will ensure it.

There are various degrees of backup to ensure continuous performance. Thus, the specification can achieve different levels of security in calling for standby equipment: two 100% units (as in aircraft ignition systems), or three 50% units, or two 75% units—and so on. It might be that it calls for items such as pumps to have installed standby spares, or that uninstalled spares should be supplied. The specification can either set down general rules of this kind or go through the individual pieces of equipment one by one.

> When piloting a light plane, before take-off but with the brakes on, you set the throttle at maximum and watch the rev counter as you switch off first one and then the other ignition system. In fact, you expect a slight fall in rpm because twin sparks give slightly better performance. So in a sense there is a slight loss of output if one system were to fail, but the plane is still safe to fly.

The important point is that if the purchaser's specification does not lay these rules down, the contractor's designer will choose the cheapest option he can in order to produce a competitive offer. It may be so close to the limit of what is asked for that it fails to meet even minor emergencies. This is where I would like to think that some kind of DBFO contract will evolve in future years, which would allow designers to avoid the kind of disasters we often see nowadays.

SUMMARY

- You never have complete or accurate data when estimating the cost of a design.
- When calculating operating costs, it's important to define the boundaries around your calculations.
- Optimization calculations based on capital and operating costs may be misleading—they need looking at critically and presented in such a way that no one is misled.
- Unless these and similar items are specified by the purchaser, the designer in a conventional competitive tender will choose the skinniest design that will just do the job—often with disastrous results.
- Utilization and reliability are vital to engineering design but often are not specified.

Chapter 25

Health and Safety Aspects of Design

ENGINEERING AND RISK

Earlier on I defined a project as something that hasn't been done before. This means that there is always risk of unexpected things going wrong: But risk is balanced against benefit. There are risks that the contract will cost more than the estimate, in which case it might take longer than anticipated for the benefit to be realized. The Channel Tunnel was so much over budget, and the owners had to borrow so much more money, that the annual income from the tunnel is not even enough to pay the interest charges. Even in this sort of extreme example, the risks are measurable in terms of money. But what about the possibility of injuries or death during construction and operation or of damage to the environment. Can these really be measured in monetary terms?

Engineering and, particularly, construction is a risk business. Balancing risk against benefit is a matter of ethics and this chapter isn't about ethics (see Chapter 28). But it's worth noting that all the engineering institutions stress that engineers have an ethical responsibility to ensure, as far as is reasonably practicable, that their activities do not create a hazard to the health and safety of workers or to the environment. It's the *as far as is reasonably practicable* that gives us the problem.

ACCIDENTS WILL HAPPEN

My enormously long experience tells me that accidents will happen. No matter how careful people are, mistakes are made and injuries or fatalities occur from time to time. Lawyers, however, don't accept mistakes and look for someone to blame or, at any rate, to sue.

Engineering Money: Financial Fundamentals for Engineers, by Richard Hill and George Solt
Copyright © 2010 John Wiley & Sons, Inc.

Accidents in the construction industry always get media coverage, largely because it operates very much in the public eye.

In 1845 work started on the Bramhope Tunnel on the Leeds–Thirsk railway. The project employed 2000 men over a period of 4 years during which time there were 24 fatalities. That's 300 fatalities per 100,000 person-years. There are two surprising things here: first, that there were so few casualties, in spite of using new and untried technology, and, second, that any records were kept.

For the year 1992/3 the UK construction industry recorded 5.7 fatalities per 100,000 man-years. That was about a third of all industrial fatal accidents and, given the nature of engineering projects, you could argue that that's not a bad record.

SAFETY LAW

A lot of the risks in engineering projects are contractual risks—budget, program, damages, and so on—and these are subject to civil law, which covers disputes between parties. Health and safety issues are different: In most jurisdictions they are subject to criminal law. If you don't comply with health and safety legislation, you have committed a crime and could be prosecuted. If there is a fatality involved, you could be prosecuted for manslaughter. In addition you could be sued for negligence and/or breach of contract under civil law.

The most important legislation in the UK is the Health and Safety at Work Act of 1974, which covers all aspects of health and safety in every kind of working environment from construction sites to shops and from garage workshops to agriculture. In plain words, the act states that an employer has a statutory duty to care for the health, safety, and welfare of employees and any other people who may be affected by their activities (e.g., the employees of contractors or members of the public). The act also set up the

---

In my 20 years as company director of two companies we had only one serious accident. One of our commissioning engineers was found dead on the floor of the plant house at the foot of a 1.5-m high step ladder. He had obviously climbed the ladder to operate a valve, fallen off, and broken his skull on the concrete floor. There was no sign of carelessness, recklessness, or drunkenness. "Can't understand it," all his colleagues said, "John was always so careful. He'd never climb a ladder while holding a notebook or anything like that." Privately I wondered whether that wasn't the problem. He had no practice at falling, so when he did have his first fall (as sooner or later everyone will) he didn't know how to protect himself so it was also his last.

Safety records from Brunel's Rotherhithe Tunnel under the River Thames (1825–1843) were so detailed that we know that the first fatality occurred on July 12, 1825, when an old ganger by the name of Painter had been out drinking, after finishing his shift in the tunnel. On his way home, he climbed the shaft. Due to his drunken state, he lost his balance and fell to the bottom of the 42-ft shaft and was killed.

health and safety executive to implement the legislation. The Health and Safety at Work Act made the contractor responsible for safety on construction sites. Similar legislation exists in other jurisdictions.

In 1994 the European Construction Industry Directive came into force. These regulations place a legal obligation on everyone involved in the construction process to provide for site safety at every stage of a project. The contractor no longer has sole responsibility, but clients and other construction professionals such as designers, consulting engineers, project managers, surveyors, architects, and landscape architects all share in that responsibility.

This is not the place to discuss the details of the Construction Industry Directive, but it places responsibility on the designer for safety during the construction phase of the project as well as in future operation. Whether the directive is entirely responsible is questionable, but there has been a steady decline in fatalities in the construction industry, which were 4 per 100,000 person-years in 2002–2003.

RISK ASSESSMENT

Florida is one of 26 U.S. states in which cities are not required to comply with federal Occupational Safety and Health Administration (OSHA) rules. The Florida Legislature dissolved the state's Division of Safety in 2000, when Govenor Jeb Bush directed all state agencies to review OSHA regulations, but he made compliance voluntary. In January 2006 two workers at the Bethune Point Wastewater Treatment Plant were using a cutting torch on a damaged methanol storage tank. They had had no training nor were there any safety procedures in place. The cutting torch ignited the vapor and the resulting explosion killed both men.

A company director received a suspended prison sentence following his conviction for manslaughter as a result of a fatal accident at a factory in Lancashire in 1988.

On May 23, 1984, a group of 44 people was taken on a tour of a water pumping station at Abbeystead. While in the valve house an explosion occurred in which 16 people died. The explosion was caused by the ignition of accumulated methane. On the evidence that there was a risk of methane being present that should have been taken into account in the design, the designers (Binnie & Ptrs, consulting engineers) were held liable while the contractor (Nuttalls) and the operator (North West Water) were held not liable on appeal.

We frequently hear that human life is infinitely precious. This is true for the individual and his nearest and dearest but can hardly be considered true for society as a whole. In fact, there are various examples in which the value of a human life must have a number attached to it. For example, how much does the designer of an ocean liner spend on lifeboats and other rescue equipment? Who decides on that figure, and how is it arrived at?

If we are going to place a cash value on life then, logically, that cost should be the loss that we would incur if we were to be held responsible by a court for the death. This is likely to be the sum paid out by insurance companies on accidental death (currently in the UK this is about £100,000–£300,000).

COST OF SAFETY

The purchaser may well recognize the legal and moral needs for health, safety, and environmental aspects of the project but will, not unreasonably, ask "How much will it cost?" If the safety input is timed correctly, the cost can be minimal. Indeed taking account of health and safety during design can avoid costly prob-

On December 3, 1984, the Union Carbide plant in Bhopal, India, accidentally released methyl isocyanate gas. The government of Madhya Pradesh confirmed a total of 3787 deaths related to the gas release and 25,000 have since died from gas-related diseases. Twenty-five years on, 390 tons of toxic chemicals abandoned at the plant continue to pollute the groundwater in the region, affecting thousands of Bhopal residents who depend on it. There are currently civil and criminal cases related to the disaster ongoing in the U.S. District Court, Manhattan, and the District Court of Bhopal, India, against Union Carbide, now owned by Dow Chemical Company.

lems later and actually reduce project costs in both construction and maintenance. Safety is usually only expensive if it is introduced at too late a stage. If managed correctly, safety can be inexpensive and cost effective. Accidents invariably result in delays and therefore large costs. Good safety can be a sound investment if it is carried out as an integral part of the project.

A bigger concern is competitive tendering. The contractor may see the need for safety in the design but will be wary of the cost. "What if my competitor doesn't include the same level of safety?" he will say, "His price will be lower than mine and I'll lose the contract." The thrust of the European Construction Industry Directive is to make all parties including, most importantly, the purchaser, aware of health, safety, and environmental aspects. This means that safety should be fully covered in the specification so that provided all the tenderers comply with the specification the playing field should be level.

SUMMARY

- **In most jurisdictions, health and safety is covered by legislation under criminal law.**
- **Compliance with health and safety legislation represents a cost to a contractor.**
- **It is difficult to assess health and safety risks in cost terms.**
- **Accidents in construction are expensive and cause delays.**

Chapter 26

Green Engineering and Greenbacks

THE ENVIRONMENT

The environment is everybody's business but, in the fight to reduce greenhouse gas emissions, we engineers are in the front line. This fight has two fronts—to use fossil fuels more efficiently or replace them (by, e.g., wind power). All uses of fossil fuels involve engineering and so do most alternatives to them.

Of course, we could just use less power but, though thrift is simple and trouble free, it has gone out of fashion. It seems we can't do without all-night street lighting, overheated homes and swimming pools, and all other energy-guzzling appurtenances of modern life.

Money is, of course, central here, especially now that there are charges for carbon emission. This book started by saying that money measures everything that engineers need. If that were truly the case, the cheapest solution to any problem would also be the greenest, but money is a bad measurement, and very bad at costing the environment. Thus, the sensible thing is to find the cheapest solution to any problem first and then see whether (and where) that fails. In particular, look doubtfully at anything that depends on government subsidy (hidden or otherwise) to make it economical. Quite likely it's not really the right solution.

We now have charges on carbon emissions, to cost that aspect of the environment, though its accuracy is doubtful, and so is its effectiveness in practice. (A tax on energy use might be more effective because of its direct impact on every citizen.) I recommend *Costing the Earth* by Frances Cairncross (Economist Business Books, London, 1991), which claims that all governments' environmental initiatives have made things worse rather than better and provides good evidence for the claim.

Engineering Money: Financial Fundamentals for Engineers, by Richard Hill and George Solt
Copyright © 2010 John Wiley & Sons, Inc.

Consider a more recent example: filament lightbulbs have now become unobtainable throughout the European Union (EU). Only the new "economy" bulbs are on sale.

In my wine cellar I have a 40-W filament bulb. As a steady drinker I need to go down and fetch a dozen bottles, say, 28 times a year. When I do that, the light is on for, say, 15 minutes. On this conservative estimate, it consumes 0.28 kWh per year, and the cost leaves change from 5 ¢/year. Not much scope for savings, then.

There must be hundreds of thousands of filament bulbs working in such conditions. The economy bulb, which will compulsorily replace them, costs more, its manufacture generates more carbon dioxide, and its mercury content leads to disposal difficulties. All counterproductive.

I served on a national environmental panel in the 1970s and have worked in the field since then. One of the most important lessons I learned on the way was on my first day on this panel, when a fellow member said "Beware of SIFs!" What's an SIF? I asked. "It stands for Single Issue Fanatic" he said. "They're people who concentrate on one specific issue and are oblivious of the massive damage they might do elsewhere." There are lots of them, and they create problems usually through ignorance or lack of thought, rather than fanaticism.

> A good engineer might combine money and technology to reach a better system. Filament bulbs could be made to operate at higher temperatures, when they use a little less power for the same light output, but have a shorter life. Suppose we replace today's 40-W filament bulb with a cheap one giving the same light at 30 W, and that burns out after 100 hours? People would soon learn which kind of bulb to use where. It's all about utilization (see Chapter 24).

What about electric cars, for example. Recharge them off the mains overnight! Clean, noiseless cities! No CO_2 emission! Wonderful!

No, it's not. Unless the power station is "green," it just moves the problem out of sight. In fact it might make it worse because electric motors may use more power if you add the losses in transmission, rectification, and storage.

Environmental problems are so varied that making generalizations is not useful. They can also be horribly complex. This is a field in which it is even more important to have the real objectives fully set out. That's where politicians tend to fail. They mostly know and care far less about the environment than about the next election. It's a good start to look at the financial costs of all alternative actions, and go on to ask where money fails to be a useful tool.

RISK

Risk to the environment is just as difficult to quantify. Costs resulting from environmental damage during construction would be the responsibility of the purchaser and are likely to be passed on to the contractor under consequential loss. The loss of amenity value and income in a trout fishery destroyed by chemicals can be reasonably estimated, as can the cost of cleaning it up and restocking it at the polluter's expense. On the other hand an established tree felled by mistake cannot be reinstated, a rare orchid destroyed by a bulldozer cannot be replanted, and a listed building demolished in error cannot be rebuilt exactly as it was. The envi-

> Veolia Water constructed a filtration plant at a water works site, which is rather special as it is home to both Greater Crested and Smooth newts and the contract required the contractor to ensure that no harm came to the newts. All site staff had to be instructed on how to handle the newts, and any that strayed into the construction area had to be relocated to the adjacent environmental center.

ronmental costs are unquantifiable and, in any case, a matter of opinion. Some people are passionate about orchids while others are completely unmoved. Nevertheless, the world is shocked by environmental disasters of the magnitude of the *Exxon Valdez* oil spill or the 2010 BP oil leak in the Gulf of Mexico.

SUSTAINABILITY

Sustainable development is a term first used by the United Nations' 1983 World Commission on Environment and Development (the Brundtland Commission) and is generally defined as development that *meets the needs of the present without compromising the ability of future generations to meet their own needs.* In other words it means that when we engineers take on a project we need to not only minimize its environmental impact in construction but that ongoing operation should minimize the use of natural resources. So it's about using renewable energy, reducing waste, recycling materials, and reusing resources.

With rising costs of fossil fuels, it makes sound economic sense to minimize energy requirements. Similarly, as the cost of waste disposal increases, recycling and reuse becomes cost effective. So there is no inherent conflict between low costs and sustainability. The provision of the facilities needed for recycling and reuse is likely to increase the capital cost of a project but will normally reduce the operating costs, so that the increased capital cost will give a rapid payback.

SUMMARY

- **Money is particularly bad at measuring environmental matters.**
- **In spite of that, choosing the cheapest solution to a problem is a good starting point.**
- **Environmental issues are always complicated, and every ramification needs to be considered.**
- **Thrift is good—and should be sustainable.**

Chapter 27

Research and Development

If scientists and technologists had more say in running the world, it would not have become normal to clap research and development (R + D) together. Actually, they are fundamentally different:

- Research is about knowledge.
- Development is about money.

Clearly, therefore, a book about money must discuss development as well as some things that we call research but (on the above definition) aren't. When engineers engage in "research," it is usually aimed at improving our understanding of something that has a practical application—that is to say, it too is ultimately about money.

Engineering companies need R&D because it introduces the new technology that will provide future profits. If they don't develop this technology themselves, they will have to license it from someone else and that adds to their costs, making them uncompetitive.

After one of his public lectures and demonstrations about electricity, Michael Faraday (1791–1867) was asked by a lady "But, Mr Faraday, what use is it?" He answered "What use, madam, is a new-born baby?" Of course, the great man was right, even if the lady, and most of the world since her, let themselves be misled by his reply. The fact is that a new-born baby is no use whatsoever. Making babies is an inexpensive and (usually) pleasurable activity, but turning the baby into a useful citizen, by contrast, takes many years of tedious and costly toil. Faraday could not have found a better way of explaining the difference between research and development.

CHALLENGE OF R&D

Young people who are choosing their career and thinking of taking up research should be clear about its true nature. It's not the romantic trade that it is cracked up to be: it rarely ends up with new and exciting discoveries. Most R&D is about failure. The thousands of doctorates published annually around

Engineering Money: Financial Fundamentals for Engineers, by Richard Hill and George Solt
Copyright © 2010 John Wiley & Sons, Inc.

the world seem to represent success, but a great number hide one or more of these facts:

> *"For me it is all about the challenge. The more difficult the task the greater the challenge. That is why I have always failed where others have succeeded."*
> —Inspector Clouseau (Peter Sellers) in *The Pink Panther*

- We reached no sensible conclusion.
- It's become clear that we set out to answer unanswerable or pointless questions.
- The experiments didn't work because ... (name any of dozens of reasons).
- We ran out of time.

This seems to make R&D an unattractive trade. It isn't really, but it needs workers who are both determined and resilient—without those qualities it is too easy to get discouraged. Quitters should do something more suited to their temperament.

That, in turn, explains why the most difficult thing for people working on an R&D project is to recognize when they've reached a point at which it is useless to go on. Good R&D workers are like mountaineers. They're strongly motivated

> This is my experience, after 45-odd years in industrial R&D and supervising postgraduates doing research, and writing it up for their theses (and 30-odd years climbing the highest peaks in the Alps).

to get to the top and expect it to be a long, hard journey. It's hard for them to say "This is no good. We have to turn back."

BACKING THE RIGHT RUNNER

How do we decide that an R&D project is worth the money it will cost to carry it through? All commercial decision making is a form of betting, when you must balance risk against reward. For decision making in engineering design, NPV, or payback, will give us some reasonable guidelines (see Appendix 3). These calculations ensure that, in betting terms, we only back odds-on favorites.

With R&D the chances of success are generally much smaller, but we hope the rewards will be much greater. In betting terms, therefore, financing R&D means backing outsiders. That, as any betting man knows, is a triumph of hope over expectation. It does, of course, need sound judgment and a good nose for opportunities, but some of it comes down to faith.

> Insurance is another branch of the gambling industry. By insuring your house against fire, for example, the insurance company is betting you that your house won't burn down this year. Like bookies at the races, the insurance companies lay off their risky bets, which means that they can't lose—but in the long term you will.

To treat R&D decisions in terms of gambling is quite reasonable. A large pharmaceuticals company, for example, is likely to have so many R&D projects running simultaneously that at least one of them ought to succeed. That one will make enough to pay for all the others' failures.

Accountants play an important part in running any kind of business—quite often even the CEO is an accountant by training. That's, of course, quite reasonable, even though engineers tend to complain about it. The money men, however, have difficulties with R&D.

One of the reasons General Motors failed in 2009 was a lack of investment in product development. CEO Rick Wagoner, who was forced to resign, was an accountant and the company's former CFO.

I have described accountants as "poets and dreamers" but that is only as far as their own field is concerned. The horizon for most accountancy work is about one year, or at most two years, so their dreams operate over short time spans. R&D—and especially D—is a long-term business, for two reasons. First, the work itself may be slow and take years. More importantly, a new invention may take far longer to become established (and profitable) than anyone expects. Accountants don't like financing projects whose profitability is beyond their horizon.

The first reverse osmosis (RO) plant for converting saline water into drinking water ran in Koalinga, California, in 1962. The next year the U.S. government sponsored a symposium on desalination, where I went to the White House and heard President Johnson forecast seawater conversion at 25¢ per 1000 gallons. Now, almost half a century later, the first three very large projects for seawater conversion by RO are nearing completion. The cost is much higher than the president predicted and is only economical because water costs from other sources have risen to make it so.

Reverse osmosis is now established as the best process for converting saline water into drinking water (except where special conditions favor distillation). It was 20 years from the time it was first shown to be practicable before it became a real commercial success. Several of the companies that promoted it have gone under since then.

This follows an all-too-common pattern: Novel innovations promise success at first, but it is in the field where snags appear and confidence sinks. The promoters lose courage or even go bust. Even with inventions good enough to recover, it all takes time. All this makes R&D a doubtful prospect to accountants, who do not naturally think in terms of very long periods. And unless the promoters have enough capital (or income from other business) to see the project through, they're right.

Since it is difficult to justify R&D in accountants' terms, it's not surprising that when business is bad and something has to be cut, accountants will normally point to R&D as the first victim. This is very wrong. For R&D to succeed, it needs the security of a long-term budget, and that of course has to be based on sensible finance. Sadly, I have only ever met one company that had a ring-fenced R&D budget for a rolling 5 years ahead. It was a French company, and their R&D was rather successful.

A smaller company may well generate excellent ideas for development but may not be big enough to finance R&D on a sensible scale. One way of saving money is to get the work done in a university. They can make it a Ph.D. student's research project and try to get a research grant for it. It often helps to get the grant if the company contributes toward the cost. The grants are usually decided by *peer review* (the opinion of academics working in similar lines of technology) to decide whether the project is worth whatever it is going to cost. In practice, this means the company must find a professor who is skilled at making a case for research and has enough chums in other universities to give it the nod.

In the 1980s Margaret Thatcher, whose economics worked well in theory but rather failed in practice, convinced a lot of UK businessmen that "downsizing" to give a "lean and mean" organization would lead to bigger profits. In particular, she suggested that R&D was wasteful and that we could simply license new technology as we needed it from other countries. It worked well in the short term and most of the UK's R&D facilities were closed. The result is that we now pay large sums in licence fees but, alas, make low profits.

In the 1990s a bunch of academic engineers at the University of Cambridge demonstrated that it was aerodynamically impossible for the bumble bee (*Bombus terrestris*) to fly. So far this does not seem to have worried the bumble bee.

There are other ways of getting grants. For example, one good source in Europe is to get into partnership with a body in some other EU country: Brussels is keen to promote such international schemes and pays handsomely toward the costs.

Commissioning R&D work as a student's project means that the prime objective of the exercise is no longer commercial usefulness, but getting the research worker his Ph.D. This puts some constraints on the work—especially with respect to time, as experimental work has to stop quite early to allow enough time for writing up the thesis.

I often hear complaints that university work tends to stray into areas that are of no interest to the company who sponsored it. This is bound to happen if the sponsor doesn't stay in close contact with the work. The sponsor's representative must be someone in a senior position, and he should visit the research facility regularly. My experience is that universities fall over themselves trying to be commercially useful. However, without regular guidance from the sponsor as the work goes on, they can't judge what is really wanted. Working with universities in this way has an important benefit—there's a wonderful lot of expertise and information in any university, and they will probably make it available free if you are sponsoring a project there. If nothing else, they may be able to tell you where to find what you are looking for.

PATENTING

Whoever has done the work, the question of patenting will come up if it promises a useful result. It raises problems that are outside the scope of this book, and they are difficult problems because full patent cover costs a lot of money. It is worth getting a good patent agent to help with them.

One drawback to commissioning work to be done by a Ph.D. student is that it must eventually be published, and therefore any patent application must predate the publication of the thesis. There are sometimes ways of delaying publication, especially when the work is of military importance. You can, of course, commission a university department to carry out your research on normal commercial lines, without their getting a student to do it for a higher degree. That shows little advantage, as it costs about as much as farming it out to a commercial research company.

HIDDEN BENEFITS OF R&D

A final word: Technologists who are good at their work and interested in it, enjoy the challenge and novelty of R&D work. Without it, they get bored. A company that doesn't supply this kind of stimulus tends to end up with mediocre staff. R&D departments also serve to train people, and they provide a reserve of people who are familiar with the company's products and technology. These benefits can't be measured in any way that an accountant would recognize, but they undoubtedly exist.

I met my co-author decades ago when he was a student working for me during his university course. He'd been with us a few months when we had a real crisis—an unhappy client who demanded an instant visit from an "expert." We hadn't anyone to send—except him. He hadn't yet learned very much but had two great virtues—he looked older than he was (*but I have worn well since*), and he knew how to bluff when he didn't know. It was the start of a great partnership.

SUMMARY

- **R&D is a gamble, usually with long odds.**
- **The most difficult thing is to recognize the point when an R&D project is failing and should be stopped.**
- **Sponsoring R&D by postgraduates working for a higher degree is cheap, but it's only useful if the sponsor stays in close contact with the work.**

Chapter 28

The Love of Money

ETHICS

Ethics is a major branch of philosophy that studies values and customs and deals with concepts such as right and wrong, good and evil, and moral standards. In other words it's about how we behave toward each other. But what do we mean by *each other*?

As law-abiding members of society we will be governed by the laws of our country, and in our day-to-day work we will obey regulatory bodies such as the Environment Protection Agency, the Health & Safety Commission, and the Food and Drug Administrate. But how do we go on when there are no guidelines?

As individuals we take on responsibilities to provide ourselves and our families with a reasonable quality of life and standard of living. For some that may be enshrined in religious beliefs. As employees we have a duty to our employer or shareholders to try to ensure that the company survives, makes a reasonable profit, and looks after its employees. As professional engineers we will have signed up to some sort of code of ethics that will define our responsibilities to clients and to society in terms of health, safety, the environment, honesty, and reasonable competence. But what do we mean when we use the word *reasonable* and what happens when our responsibilities to society, our employer, and our family come into conflict?

This book is about engineering and money, and money has always raised ethical problems.

ENGINEERING AND ETHICS

Engineers and warfare owe a lot to one another (see Chapter 23). Many of the advances in aeronautics, electronics, nuclear engineering, and many other areas were militarily driven—but how do we view the end result of our labors? Fritz Haber, the Nobel prize winning chemist, made a lot of money out of

---

Engineering Money: Financial Fundamentals for Engineers, by Richard Hill and George Solt
Copyright © 2010 John Wiley & Sons, Inc.

developing poison gas as a weapon in World War I and showed no regrets over the huge numbers of soldiers it killed.

In World War II the extermination camps of the Holocaust were designed by German engineers. The engineering drawings clearly showed the purpose of these projects. The argument is always that we might as well build it and make some money: if we don't someone else will.

A major area of concern for engineers in the construction industry these days is disposal of waste materials, particularly contaminated land and chemicals, and the potential for environmental damage. In most countries, the environmental protection laws and regulations make waste disposal expensive, and there is a great temptation for contractors to save money by disposing of wastes illegally. We have all read of major environmental disasters ranging from killing fish in a river by dumping cyanide waste to long-term health problems caused by landfill sites that have not been properly constructed and sealed.

> Alfred Nobel (1833–1896) made a fortune from developing modern explosives. When he realized that he would be largely remembered for manufacturing products that killed people, he decided to leave his fortune to found the Nobel Prize to reward outstanding work for science, peace, and the arts.

> Mikhail Kalashnikov, on the other hand, made very little money from his AK47 automatic rifle. It is a brilliant piece of mechanical engineering: few moving parts, easy to service, and low production cost. It is the most successful assault rifle in the world. In an interview in 2007 Kalashnikov said "I'm proud of my invention, but I'm sad that it is used by terrorists. I would prefer to have invented a machine that people could use and that would help farmers with their work—for example, a lawnmower."

MONEY AND ETHICS

In the words of the Emperor Vespasian, *pecunia non olet* (money does not smell). Although he was talking about his tax on urine (a vital raw material for the chemical industry of the day), he was correct in that money per se has no attributes other than the value that it represents.

On the other hand, St. Paul tells us in the Bible that *the love of money is the root of all evil* (I Timothy 6:10). And avarice is one of the seven deadly sins. While the love of

> I live in a small town in the UK called High Wycombe and every year we elect a mayor. By tradition the mayor is weighed in public before taking office and again on his last day. A mayor who puts on weight during his year in office is regarded as having lived too well at the expense of the townsfolk and faces public ridicule.

> St. Paul's words are often misquoted as "Money is the root of all evil." Money, of course, has no intrinsic properties— it's what we do with it that matters.

money is not the only root of society's problems, there is no doubt that personal and corporate greed played a major part in the global economic crisis of 2007–2008.

Of course, the goal of a company is to maximize the shareholders' investment, but we have no real definition of what constitutes a reasonable profit. Competition will normally limit the market price of any goods, including, as we saw earlier, engineering contracts. This is why monopoly businesses like utility supply can be bad for consumers. However, the profit margin depends not just on selling price but also on cost. So a contractor can make a bigger profit by cutting back on quality. As engineers we have a responsibility to ensure that our client gets fair value for money, and sometimes this can bring us into conflict with our commercial responsibilities.

Usury is lending money at an exorbitant rate of interest, and all the major religions frown upon it. But what do we mean by *exorbitant*? Earlier we reckoned about 5% above prime rate would be fair, but then we said that banks will charge whatever they can get away with.

BRIBERY

Bribery has always been rife in all branches of human endeavor. The English Prime Minister Sir Robert Walpole (1676–1745) remarked: "Every man has his price." In some cultures it is customary to provide a gift for the person who signs the contract, and sometimes it is quite a big one. The innocuous word "gift" is, in this case, a euphemism for bribe. When dealing in these countries, most western contractors appoint a local agent to represent them and to deal with such local niceties. By simply paying a large *agent's fee* they remove themselves from the deal making and so salve their consciences.

But what constitutes a bribe? In the West, many suppliers give Christmas gifts to their customers. This may be no more than a calendar or pen with the company name on it, or it may be a bottle of Scotch. But when does a business gift become a bribe?

Certainly a family holiday was seen as a bribe by the courts when a local government official in Newcastle-upon-Tyne in the UK, T. Dan Smith, placed a series of contracts with a local architect John Poulson. Smith and his family enjoyed a series of holidays in the West Indies that were paid for by the architect, and both men went to prison for corruption.

CONTRACTS AND OTHER GOALS

As I said earlier, it is not unusual for a contractor to be invited to bid for a contract that has a serious flaw in the consultant's design that will cost a lot to rectify. The options open are:

- Point out the flaw and risk upsetting the consultant and getting barred from bidding.
- Bid for the correct design and risk losing the contract on price.
- Decline to bid.
- Bid for the specified design (with or without caveats).

As we saw, most contractors will simply bid for the specified design even when they know that it is flawed. They hope that they will either be paid to put it right or at least that their company will be legally in the clear. They might also calculate that, if it does come to litigation, it will be so far in the future that everybody concerned will have moved on. On the other hand, it's the only course of action that gives you a fighting chance of winning the contract. But is it ethical?

The need to win contracts often leads engineers to ignore or cover up design faults. This happened in 1986 when the space shuttle *Challenger* was due for takeoff. A fault was found in a booster, but the manufacturers feared that if they delayed the launch they would not be approved by NASA to bid for the next shuttle contract. So they kept quiet about it. And *Challenger* exploded killing all the crew.

Sometimes it's political pressure. In the 1950s there was a race between the UK (De Havilland) and the United States (Boeing) to have the first commercial jet passenger airplane. The De Havilland Comet was first into service but soon there were accidents. The

> I was on honeymoon in Kenya in 1954 when the third De Havilland Comet blew up in midair, and they really did have to ground them then. That meant major delays on the air route to Africa, and we got three days' extra honeymoon out of it. Every cloud has a silver lining.

problem was metal fatigue, but the UK government persuaded the operating company to blame it on pilot error so that the Comet fleet would not be grounded and UK would stay ahead in the race.

WHISTLEBLOWING

In sports, referees blow a whistle when they see a foul. *Whistleblowers* are employees of a company who make public illegal or unethical practices within the company. They are not usually popular with corporate executives. Whistleblowing can have a high public profile. In 1996 Jeffery Wigand, a vice president of the tobacco company Brown & Williamson, appeared on the CBS

> In 2005 Paul Moore was head of Group Regulatory Risk at international bankers HBOS and concerned that high-pressure selling of unsecured loans was putting both customers and the bank at risk. He made his concerns known to the board and was sacked. HBOS failed in 2008 and had to be bailed out by the UK government with a loan of £37 billion.

news program *60 Minutes*. He stated that the company used additives in its cigarettes that were known to be carcinogenic and/or addictive. He was fired. His story is told in the 1999 film *The Insider*.

WHAT'S ETHICAL?

As engineers we often face the conflict between our duty to the company and its shareholders on the one hand, and to our client, or the environment, or society in general on the other. Moreover, taking the wrong option can lose us our job. My own way of facing such difficulties is to aim at the *least bad* solution. Extremists like SIFs (see Chapter 26) are convinced that there is a *best* solution and their idealism blinds them to the untold damage they might be doing.

> For years I tried (and failed) to become a Quaker. Quakers should always be truthful, but you can't get through life without the occasional lie. Many people lie simply because they haven't really thought about it but say the first thing that comes into their minds. I have got into the way of only telling a lie when it seems unavoidable and am glad to find that that isn't often. That's what I call a least bad solution.

Finding the least bad solution only works if you think clearly and honestly, a thing that frightens many people, so you may need to force yourself to do it. You can then make your decision, for better or worse. Maybe it really is the best, but usually it is wrong in some respect—there would have been no problem if that weren't so. At least you have thought as clearly as you can, and done your best. It's not a heroic doctrine, but at least it's practicable.

So what is ethical behavior—or, more to the point, when does behavior become unethical? The problem is that this is something beyond the scope of measurement, so we can't simply draw a line. Each individual has a different threshold, or perhaps Walpole was right and it's really just a question of price.

SUMMARY

- **Engineers often face ethical dilemmas in their work.**
- **Many of these arise from conflict between their various duties and therefore have no perfect solution.**
- **Money is always a temptation for the unscrupulous.**
- **Euphemisms don't change the meaning—they only serve to conceal it.**
- **Aiming for the *best* solution can be damaging, and it is usually better to aim for the *least bad* solution.**
- **This is a personal view: Feel free to disagree with it.**

Chapter **29**

Last Words

That's it.

I hope you now have a better idea of the environment in which you work. MBA courses don't feature most of this stuff. If you want to know more, you will have to study at the University of Life—which is, after all, the traditional way of learning the subject.

SUMMARY

- **Good luck!**

Engineering Money: Financial Fundamentals for Engineers, by Richard Hill and George Solt
Copyright © 2010 John Wiley & Sons, Inc.

Appendix 1

Financial Accounts

This is not a book about accounts, but accounting is all about measuring money, and it's important to know enough about accounting to understand what the accountant is telling you. It will influence the business decisions that you make as a project manager, contract manager, or CEO. Accountants talk about "liquidity" and "liquid assets." I want to continue the fluids analogy for a bit because it's an easy one for engineers to understand. Think of the company as a storage tank for water (Figure A1.1). Water flows into the tank rather like cash flows into the company—not necessarily at a constant rate (though that would be nice!) nor even continuously. The water flows out of the tank are also usually intermittent but controlled, although a leak in the tank might give a constant uncontrolled loss. The things we can measure are the volume of the tank contents, the instantaneous flow rates both in and out, and the cumulative flows over a period. Hang on to that picture.

Most legal systems around the world demand that, at the end of every financial year, the company has to prepare a *statement of financial position* (or, more commonly, *balance sheet*) and a *statement of income* (or *profit & loss account*) for public record. Since 2001, in most countries the accounts must conform to the guidelines of the International Accounting Standards Board, although the format of the presentation and the detailed headings do vary from country to country. One point of commonality is that the published accounts must show the previous year's figures side by side with the current year's so that comparisons can be made quite easily.

BALANCE SHEET

Back to the analogy. The balance sheet shows the tank contents on the last day of the financial year. It is calculated by adding up all the company's assets and subtracting all its liabilities. The water in the tank is the company's working capital. If the level in the tank falls too low, and no more can be found to top it up, then the company doesn't have enough assets to cover its liabilities. It becomes *insolvent* and may be forced into *liquidation*.

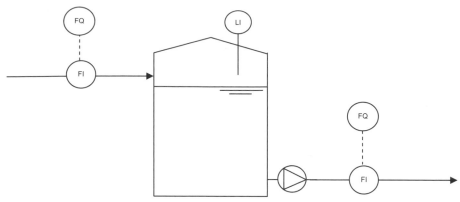

Figure A1.1   Company as a water tank.

The balance sheet is based on the fact that although the company is a legal entity, it isn't actually real so it can't own anything. Consequently, everything it has (its *assets*) must be owed to someone else (its *liabilities*) and that includes the *capital and reserves*, which is what the company owes to the shareholders. So a company's assets are always equal to its liabilities. In its simplest format, a balance sheet is shown in Figure A1.2.

Fixed assets are things that the company owns but that are not associated with the day-to-day business. Fixed assets include tangible items, such as buildings, office furniture, earth-moving equipment, and so on, and intangible items, such as shares in other companies or intellectual property such as patents and licenses. *Current assets* are directly related to the day-to-day operation. In manufacturing and retailing the principal element usually refers to *stocks*. This is short for "*stock in trade*," which means items that have been purchased either for use in manufacturing or for sale. Note that stocks in this context has nothing to do with "stocks and shares." In the case of the corner shop, stock is newspapers, magazines, cigarettes, candy, and so on that you hope to sell. This is stored in the *stockroom* and is monitored by regular *stocktaking*. In a contracting company, it means *work in progress (WIP)*; that is, equipment and materials such as pumps, tanks, control software, building materials, and services such as engineering designs that have been purchased as part of a contract. *Debtors* are those who owe money to the company and, in the contracting business, this usually means the contract purchaser. *Cash* means money that is held in the bank—ideally on deposit earning interest!

Current liabilities, like current assets, are associated with day-to-day business and include any money that has been borrowed to pay for stocks and any money that the company owes to creditors, for example, suppliers, subcontractors, employees, and taxes. *Long-term liability* is money the company must repay at some time more than 12 months in the future, which is usually identified separately and includes creditors such as long-term bank loans, provisions (allowances for things that have been identified as likely to go wrong, such as

BALANCE SHEET

	2010		2009	
	$'000	$'000	$'000	$'000
FIXED ASSETS				
Tangible assets	3712		4055	
Investments	19		19	
		3731		4074
CURRENT ASSETS				
Stocks	3044		2523	
Debtors	3544		3170	
Cash	585		2034	
		7173		7727
TOTAL ASSETS		**10904**		**11801**
CURRENT LIABILITIES				
Bank loans and overdrafts	1147		1826	
Trade and other creditors	2807		3184	
		3954		5010
CREDITORS				
Amounts falling due after more than one year	624		624	
Provisions for liabilities and charges	245		217	
		869		841
CAPITAL AND RESERVES				
Called up share capital	1183		1183	
Share Premium	95		95	
Retained profit (*from P & L*)	4803		4672	
		6081		5950
TOTAL LIABILITIES		**10904**		**11801**

Figure A1.2 Sample balance sheet.

a contract that may need remedial work, or a customer who may go bust before the company is paid—a *bad debt*). *Capital and reserves* (sometimes called *shareholders' equity*) is the money owed to the shareholders and consists of the money they originally invested together with whatever profits the company has accumulated over the years.

In the UK it is normal to present the balance sheet in a slightly different way for the convenience of accountants. The balance sheet shown in Figure A1.3 below uses the same numbers as our previous example but is presented as you would find it in a set of UK accounts. If you work through it, you'll be able to see from where the numbers have come.

BALANCE SHEET

		2010		2009	
		$'000	$'000	$'000	$'000
FIXED ASSETS					
Tangible assets		3712		4055	
Investments		19		19	
	(A)		3731		4074
CURRENT ASSETS					
Stocks		3044		2523	
Debtors		3544		3170	
Cash		585		2034	
	(B)	7173		7727	
CURRENT LIABILITIES					
Bank loans and overdrafts		1147		1826	
Trade and other creditors		2807		3184	
	(C)	3954		5010	
NET CURRENT ASSETS	(D = B − C)		3219		2717
TOTAL ASSETS LESS CURRENT LIABILITIES	(E = A + D)		6950		6791
CREDITORS					
Amounts falling due after more than one year	(F)		624		624
Provisions for liabilities and charges			245		217
	(G = E − F)		6081		5950
CAPITAL AND RESERVES					
Called up share capital			1183		1183
Share Premium			95		95
Retained profit (*from P & L*)			4803		4672
	H = G		6081		5950

Figure A1.3   Sample balance sheet for UK accounts.

The value G here must be equal to all the money that the shareholders have invested in the company and is, therefore, the same amount as the capital and reserves (H), which we talked about earlier. This is, in principle, the value of the company. Auditors always manage to get these totals to be the same, which is absolutely bogus because many of the figures that make up items A, B, and C contain a lot of guesswork. By and large the capital and reserves is a calculating convenience and of little practical importance.

Because the company's total assets and total liabilities must always be the same, numbers move within the balance sheet. For example, when we purchase materials or services for a contract, the suppliers supply the materials or services and send us invoices, which ask us to pay them. The value of the company's current assets (stocks) increases by the value of the goods or services, and, because we have invoices from suppliers, the current liabilities (creditors) increases by the same amount. When we pay the suppliers' invoices by sending them a check, the value of current liabilities (creditors) decreases and the value of current assets (cash) decreases by the same amount. So the balance sheet always balances.

DEPRECIATION

If we buy equipment (e.g., earth-moving equipment or trucks) that will be used on many different contracts and will, we hope, last for several years, then these items are regarded as fixed tangible assets. When we buy them, the value of fixed assets increases and the balance is maintained by decreasing the value of the cash assets. How-

Suppose we buy an automobile for $12,000 and expect to sell it after 3 years for $6000; the annual depreciation is $(12,000 - 6000)/3 = \$2000$ per annum. In the first year the asset value would appear as $12,000, in the second year it would be $10,000, in the third $8000, and in the fourth it would be sold for $6000 to become a cash asset.

ever, if the equipment is going to last for, say, 10 years, then putting the whole value of the asset onto the first year balance sheet gives a distorted picture. In addition, the equipment will wear out so, after 10 years, it will not be worth as much as it was when it was purchased. If you own an automobile, you will be familiar with this. To get round this problem, we show the value of the asset at its depreciated value. This is usually calculated using a straight-line depreciation method. The annual depreciation is calculated by subtracting the residual value (what we expect to sell the equipment for when we have finished with it) divided by the expected period for which we will own it.

In practice, equipment usually loses more value in the first year and depreciation is not truly linear. Accountants have various conventions for this, but the principle is exactly the same.

PROFIT AND LOSS

The second document in the audited accounts is the profit and loss account, and this gives the cumulative flows into and out of the tank over the 12 months of the financial year. The flow into the tank is the company's *turnover*, that's the total amount of money that has been invoiced during the year. A contractor's profit and loss account might also show other sources of income—for example, income from shares in a consortium or joint venture or fees from technology licenses—but these will normally be small by comparison with the income from contracts.

Flows out are represented by the *cost of sales*—that's the direct costs of the business—and by the *net operating expenses*, which is the indirect costs or overhead. Direct costs are the costs that are directly due to carrying out the company's work, which will produce the turnover. The test whether a particular cost is direct or not is to ask whether the company would have incurred this particular cost if it hadn't executed this particular contract. Materials, engineering, site cabins, craneage, extra employees, and subcontracts arising out of the work are obviously direct costs, but on more doubtful details the company's accountancy system defines what is and what isn't a direct cost. It follows that direct costs can always be attributed to a specific contract.

Overhead expenses are costs that can't be attributed to any contract. They include such things as postage, stationery, the accounts department, IT systems, or the receptionist's salary.

The profit and loss account takes the turnover and subtracts the cost of sales to give a figure called the *gross profit*, to which each contract the company undertakes makes a *contribution* (see Chapter 7). Subtracting the net operating expenses from the gross profit leaves the *net profit*, and this is the amount of money on which the company has to pay tax. After taxes have been paid it's up to the directors to decide what to do with the surplus. If the net profit figure is negative, then the company has made a *loss* for the year. That isn't necessarily a problem, provided that there is still a reasonable amount of water in the tank. The profit and loss account is set out as shown in Figure A1.4.

Note that the retained profit from last year appears on this year's figures to give the total retained profit, which then appears on this year's balance sheet. Assuming that the company makes a net profit, the directors have to decide how much of it (if any) to distribute to the shareholders as dividend and how much to reinvest in the company.

So we've looked at the tank contents and the cumulative flows over the year that have produced it. The balance sheet and profit and loss account show last year's results so that we can compare them. If the in-flow over the year has been more than the out-flow, the contents will have increased and vice versa. But, remember that the balance sheet is a still photograph of the company at an arbitrary moment in time, while the profit and loss account is a

> The share price can be a very important factor in creating confidence in a company and can be manipulated by "creative accountancy," as Enron found just before its collapse. This is outside the scope of this book. If I knew how to do it, I'd have made my millions and retired—or be in prison!

summary of the company's operations over an arbitrary period of time (normally 12 months). The audit thus gives us a two-dimensional view of the company. Just as in engineering drawings, we all know that two views aren't enough: a third dimension is needed for a really definitive picture. In the case

PROFIT AND LOSS ACCOUNT

	2010 $'000	2009 $'000
TURNOVER	25658	25323
Cost of Sales	−18967	−18582
GROSS PROFIT	6691	6741
Net Operating Expenses	−5624	−5223
OPERATING PROFIT/(LOSS)	1067	1518
Interest receivable	100	76
Interest payable	−242	−314
PROFIT/(LOSS) BEFORE TAXATION	925	1280
Tax on profit	−370	−416
PROFIT/(LOSS) AFTER TAXATION	555	864
Dividends	−424	−431
	131	433
Retained profit brought forward	4672	4239
Retained profit carried forward	4803	4672

Figure A1.4 Sample profit and loss account.

of accountancy that third dimension is time, so we need to see some years' worth of audits to get it.

Important though these measurements are, they don't tell us anything about the situation on a day-to-day basis. The cumulative flows might be enough to produce a profit at the end of the year, but if the tank gets empty—or almost as bad, overflows—at times, then, there will be problems. We need to look at the instantaneous flows in and out of the tank to see if the level is going up or down and whether it's getting near to the bottom or about to overflow. This is what management accounts do.

Appendix 2

Critical Path Analysis

WHAT'S A CRITICAL PATH?

Any and every project is made up of a number of activities, some of which depend on others being completed before they can start. We saw how, when building a house, the foundations have to be finished before the walls can be started and the roof can't be put on until the walls are up, but digging the rain water soakaway can be done at any time.

The *critical path* is that sequence of interdependent activities that together determine the duration of the project. Activities on the critical path do not have any time between completing one activity and starting the next. Identifying that critical path determines the length of the project, and each activity on it must be completed in the allocated period of time otherwise the whole project will overrun. Activities not on the critical path can overrun within a period of time, called the "float," which does not affect any critical activity.

Critical path analysis (CPA) is one of a number of tools—all now available as software packages—that have been developed to allow the project manager to calculate the time required for completion of the project and to see which are the critical activities.

ACTIVITY LIST

The first step in CPA is to prepare a list of all the activities that make up the project and, for each, identify the duration and dependent activities, that is, activities that have to be completed before another can start. Let's use the house building example as shown in Table A2.1.

NETWORK DIAGRAM

We can now draw a network of arrows and nodes (Figure A2.1)—the so-called *activity on arrow* diagram—in which the arrows represent the activities (each

Engineering Money: Financial Fundamentals for Engineers, by Richard Hill and George Solt
Copyright © 2010 John Wiley & Sons, Inc.

Table A2.1

Activity	Description	Duration (Weeks)	Dependent (Activities)
A	Access road	3	
B	Footings	2	
C	Foundations	2	A, B
D	External walls	6	C
E	Internal walls (lower)	4	C
F	Floors	2	E
G	Internal walls (upper)	3	F
H	Roof trusses	2	D
J	Roof tiles	2	H
K	Gutters/drain pipes	1	J
L	Soakaway	1	

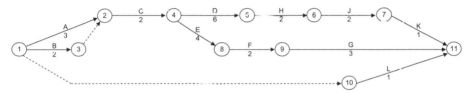

Figure A2.1   Activity on arrow diagram.

identified with the duration marked on it) and the *nodes* (sometimes called *events*) are the start or completion of an activity. This shows how the activities depend on one another The dotted line connecting node 3 to node 2 is a *dummy activity*. This shows dependence and means that activity C cannot start until activities A and B have both been completed. Dummy activities have no duration.

We can now assign to each activity an *earliest start time* (EST) by looking at the longest activity path that precedes it. That will take us to the earliest time at which we can reach node 10, which is completion of the project (in this case 16 weeks). So:

EST for activity C = EST for activity A + duration of activity A = 0 + 3 = 3.
EST for activity D = EST for activity C + duration of activity C = 3 + 2 = 5.
EST for activity E = EST for activity C + duration of activity C = 3 + 2 = 5.
EST for activity F = EST for activity E + duration of activity E = 5 + 4 = 9.
EST for activity G = EST for activity F + duration of activity F = 9 + 2 = 11
    and so on.

Doing the same thing in reverse, starting at node 10 and working backward, gives us the *latest start time* (LST) for each activity. So:

Table A2.2   Critical path Data

Activity	Duration	EST	LST	Float
A	3	0	0	0
B	2	0	1	1
C	2	3	3	0
D	6	5	5	0
E	4	5	7	2
F	2	9	11	2
G	3	11	13	2
H	2	11	11	0
J	2	13	13	0
K	1	15	15	0
L	1	0	15	15

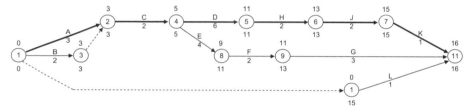

Figure A2.2   CPA diagram.

LST for activity G = 16 – duration of activity C = 16 – 3 = 13.
LST for activity F = LST for activity C – duration of activity F = 13 – 2 = 11.
LST for activity F = LST for activity F – duration of activity E = 11 – 4 = 7
    and so on.

The difference between the latest start time and the earliest start time for each activity is the *float*. The path with zero float is the critical path (see Table A2.2).

We can now mark on each node the earliest start time and latest start time for the following activity and also highlight the critical path (Figure A2.2).

The numbers above the nodes are called the *earliest event times* (EET) and represent the earliest starting time for the start of the next activity. The numbers below are the *latest event times* (LET) and represent the latest possible start time for the next activity that will achieve the project time, which is the sum of the times on the critical path and is shown on the last node. The nodes on the critical path always have the same earliest event time (EET) and latest event time (LET). It is usual practice to include this information in the node as shown in Figure A2.3.

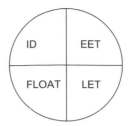

Figure A2.3   Node symbol.

EST	LST
ACTIVITY DESCRIPTION	
DURATION	FLOAT

Figure A2.4   Activity on node symbol.

Note that this is only one approach (and the simplest) to setting out critical path analysis, and there are several different ways of presenting the network diagram and of annotating it with durations, times, and float. In some versions, called either *activity on node* diagrams or *precedence* diagrams, the nodes represent activities and the arrows or lines represent the relationships of activities and their precedence. The convention in this system is to use a rectangular node like that shown in Figure A2.4.

Appendix **3**

Project Evaluation Techniques

TO BE OR NOT TO BE ...

We are now going to took at the way the client for a project might first decide whether it is sensible to go ahead with it—that is, whether the prospective benefit is worth spending a lot of money on development, conceptual design, and, eventually, construction. The hope is that the project will bring in an income, which is usually measured in terms of money, and that this will more than cover the costs and so provide a profit. The income might be from sales of a product—perhaps a pharmaceutical, a semiconductor device, or a petrochemical—or from rent on a building or tolls from a bridge. The same logic applies no matter what the undertaking. How does anyone determine whether it's worth making the investment?

PROJECT CASH FLOW

Let's consider a hypothetical project to manufacture a new plastic. The R&D department reckons that it will take a year to develop the synthesis route in the laboratory and carry out pilot-scale trials to make sure that the manufacturing process will work, and that that will cost about $5 million. Engineering has talked to contractors and estimates that the cost of constructing a full-scale manufacturing plant will be about $40 million over 2 years—$25 million in the first year and $15 million in the second year. The marketing department is optimistic about the new product and reckons that it will be able to sell $5 million worth in the first year, double that in the second year, and then sell $20 million a year for the next 4 years. After this, competitive products will come onto the market, so sales will probably fall to nothing over the next 4 years.

Engineering Money: Financial Fundamentals for Engineers, by Richard Hill and George Solt
Copyright © 2010 John Wiley & Sons, Inc.

178

Table A3.1   Cumulative Cash Flow

Year	Income ($ 000)	Cumulative ($ 000)
0	−5,000	−5,000
1	−25,000	−30,000
2	−15,000	−45,000
3	5,000	−40,000
4	10,000	−30,000
5	20,000	−10,000
6	20,000	10,000
7	20,000	30,000
8	20,000	50,000
9	15,000	65,000
10	10,000	75,000
11	5,000	80,000
12	1,000	81,000

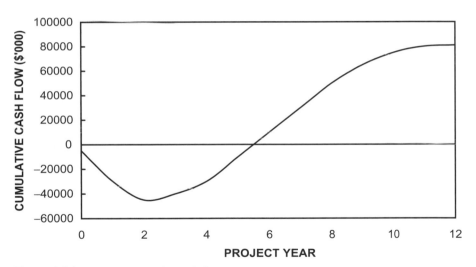

Figure A3.1   Project cumulative cash flow.

Table A3.1 shows the cumulative cash flow for the project, and this is shown graphically in Figure A3.1.

Thus, over the 12-year anticipated life of the project, the costs will be $45 million and the total income will be $126 million. That is a gross profit of $81 million, which represents a return on investment of 81/45 = 180% over 12 years or 15% per annum. Or does it?

Let's look at some other methods of project evaluation.

PAYBACK

If a capital sum $P is invested in a project that brings in an income (or effects a saving) of $S per annum, then the *payback period* of the project, T years, is P/S. In year $T + 1$ the project starts making a profit. The shorter the payback period, the better the investment.

In our project the payback period, when the cumulative cash flow curve crosses the x axis, is about 5.5 years.

It's actually a bit more complicated than that for a number of reasons. First, big capital projects rarely bring in the same income every year throughout their life. Second, most companies have to borrow the capital to invest. If a company borrows a capital sum $P at an interest rate of i and repays it after T years, the total cost is

$$P_T = P(1+i)^T$$

So a sum of $45 million borrowed at 5.5% per annum interest would result in a debt due after 5.5 years of

$$45 \times 1.055^{5.5} = \$ 60.4 \, \text{million}$$

This increases the payback time, which, in turn, increases the cost because of the additional interest incurred.

AMORTIZATION

Amortization means converting a capital debt into annual repayments. If a company borrows a capital sum $P over a period of T years, then, as a first approximation, we can assume that the debt is paid off linearly with time, so that the rate of amortization, $S per annum, is

$$S \cong P/T$$

If we borrow $45 million and pay it back over 5.5 years, then the annual payments is

$$S \cong 45/5.5 = \$8.2 \, \text{million per annum}$$

In fact, we have to pay interest on the money—remember we said that if we use our own money, then it's still costing us interest, which is the interest we would have earned from it if we hadn't spent it. If the company borrows the money at an interest rate of i and repays it over T years, the annual cost is

$$S = \frac{iP}{1-(1+i)^{-T}}$$

so our \$45 million 5.5% interest over 5.5 years would cost

$$\frac{0.055 \times 45}{\left(1 - 1.055^{5.5}\right)} = 9.7 \text{ million per annum}$$

Converting capital cost into an annual cost allows us to build up a total annual cost, including direct operating costs and overhead, which is useful for comparing different options. But it doesn't take any account of how the value of money changes with time.

TIME VALUE OF MONEY

The value of a sum of money \$P at some time T years in the future at an interest rate of i is given by

$$P_T = P(1+i)^T$$

So the *present value* (PV) (the value in today's money) of \P_T in T years time is

$$P = P_T (1+i)^{-T}$$

If we look at our steady income of \$20 million per annum over 4 years and assume a rate of 5.5%, by today's standard it's not worth the \$80 million we originally assumed; it's only worth \$70.1 million as the following table shows:

Year	PV (million dollars)
1	18.96
2	17.97
3	17.03
4	16.14
Total	70.10

This calculation of the present value of money that we will have in the future is called *discounting*.

NET PRESENT VALUE

The *net present value* of a project takes our original cash flow calculation and discounts the future income to today's value so that it is directly comparable to the capital investment, which is also at today's value. The net present value of the project is

$$NPV = PV \text{ of future income} - \text{Investment cost}$$

So, if the NPV is greater than zero, the project will do better than breakeven.

RATE OF RETURN

A slightly more sophisticated approach allows comparison between different investment opportunities. The discounted cash flow internal rate of return (DCF IRR) is the interest rate, which gives an NPV for the investment of zero. The calculation is not easy and the simplest approach is to use trial and error or the goal-seeking function on a spreadsheet. Actually, Excel has a built-in DCF function—if you can figure out how it works!

The following table compares the cash flow forecasts for three different projects. In each case the capital investment is the same at $45 million and in each case the total revenue is $81 million over 12 years, but the revenue forecasts are quite different.

Project A is the novel plastic example we've been considering. Project B gives a higher peak income but is slower to reach it, while project C gives a lower peak income but one that is more even through the life of the project.

	Project A			Project B			Project C		
Year	Income ($ 000)	PV ($ 000)	Cum ($ 000)	Income ($ 000)	PV ($ 000)	Cum ($ 000)	Income ($ 000)	PV ($ 000)	Cum ($ 000)
0	−5,000	−5,000	−5,000	−5,000	−5,000	−5,000	−5,000	−5,000	−5,000
1	−15,000	−14,218	−19,218	−15,000	−14,218	−19,218	−15,000	−14,218	−19,218
2	−25,000	−22,461	−41,679	−25,000	−22,461	−41,679	−25,000	−22,461	−41,679
3	5,000	4,258	−37,421	500	426	−41,254	15,000	12,774	−28,905
4	10,000	8,072	−29,349	1,500	1,211	−40,043	15,000	12,108	−16,797
5	20,000	15,303	−14,046	5,000	3,826	−36,217	15,000	11,477	−5,320
6	20,000	14,505	459	10,000	7,252	−28,965	15,000	10,879	5,559
7	20,000	13,749	14,207	25,000	17,186	−11,779	15,000	10,312	15,870
8	20,000	13,032	27,239	25,000	16,290	4,511	15,000	9,774	25,644
9	15,000	9,264	36,504	25,000	15,441	19,952	15,000	9,264	34,909
10	10,000	5,854	42,358	20,000	11,709	31,661	15,000	8,781	43,690
11	5,000	2,775	45,133	10,000	5,549	37,210	5,000	2,775	46,465
12	1,000	526	45,659	4,000	2,104	39,314	1,000	526	46,991

The following table compares the profitability of the projects.

	A	B	C
Capital cost, $ 000	45,000	45,000	45,000
Σ Income, $ 000	126,000	126,000	126,000
Profit	81,000	81,000	81,000
Annual return	15%	15%	15%
Payback, years	5.5	7.5	5.5
PV of income	87,338	80,993	88,670
NPV at 5.5%pa, $ 000	42,338	35,993	43,670
DCF IRR	16.8%	13.5%	17.9%

This table shows that if we calculate the profit for each project as we did at the start of this appendix, then they all show the same $81 million profit (15% per annum). On a simple payback calculation, the best option is either A or C, both with 5.5 years payback. Using NPV as a criterion puts C just ahead, although the difference becomes smaller if we take a lower value for i. The DCF IRR makes C a clear winner, and this result is independent of i.

This example is really quite unrealistic because projects that are being compared will have different investment costs and different revenues, but it illustrates the technique and also shows just how important the timing of income can be.

The following table shows a cash flow projection for three completely different projects with different capital costs ($40 million, $100 million and $20 million) and different incomes.

	Project D			Project E			Project F		
Year	Income ($ 000)	PV ($ 000)	Cum ($ 000)	Income ($ 000)	PV ($ 000)	Cum ($ 000)	Income ($ 000)	PV ($ 000)	Cum ($ 000)
0	−40,000	−40,000	−40,000	−100,000	−100,000	−100,000	−20,000	−20,000	−20,000
1	1,000	948	−39,052	5,000	4,739	−95,261	1,000	948	−19,052
2	2,000	1,797	−37,255	10,000	8,985	−86,276	5,000	4,492	−14,560
3	5,000	4,258	−32,997	40,000	34,065	−52,212	10,000	8,516	−6,044
4	10,000	8,072	−24,925	40,000	32,289	−19,923	10,000	8,072	2,028
5	20,000	15,303	−9,622	40,000	30,605	10,682	10,000	7,651	9,680
6	20,000	14,505	4,883	40,000	29,010	39,692	10,000	7,252	16,932
7	20,000	13,749	18,631	40,000	27,497	67,190	10,000	6,874	23,807
8	20,000	13,032	31,663	40,000	26,064	93,254	10,000	6,516	30,323
9	15,000	9,264	40,928	40,000	24,705	117,959	10,000	6,176	36,499
10	10,000	5,854	46,782	40,000	23,417	141,376	10,000	5,854	42,353
11	5,000	2,775	49,557	40,000	22,196	163,573	10,000	5,549	47,902
12	1,000	526	50,083	1,000	526	164,099	1,000	526	48,428

As the following table shows, project E has the highest NPV but project F shows a higher DCF IRR. Coupled with it's shorter payback project F shows a better return on investment.

	D	E	F
Capital ($ 000)	40,000	100,000	20,000
Income	129,000	376,000	97,000
Annual rate	27%	31%	40%
Payback	5.5	6.5	3.5
NPV	50,083	164,099	48,428
DCF IRR	20%	25%	32%

Index

Accounts, audited 18, 28
Accounts, financial 17, 20, 42, 102, 104, 167–173
Accounts, management 20, 28, 79, 81, 100, 110, 173
Amortization 180
Annual General Meeting 28
Articles of Association 24
Assets 18, 20, 25, 35, 102, 109, 126, 167

Balance Sheet 18, 29, 103, 126, 167
Bank of England 9, 13
Bank, Islamic 14, 59
Bank, merchant 14, 27, 36
Bill of Lading 108
Bills of materials 63
Bills of quantities 67
Breakeven 42, 182
Brunel 3, 131

Capital, fixed 34–39, 102, 167, 178–184
Capital, nomonal 24
Capital, working 34–39, 42, 53, 106, 125, 133, 167–173
Captal cost 102–105, 133, 140–142, 154, 178–184
Cash flow 49–58, 106–110, 119–123, 127, 167, 178–184
Consequential loss 50–55, 71, 79, 82–88, 94, 99, 110
Consideration 60
Consultant 61–64, 74–76, 130–137, 162
Contract Manager 6, 49, 66, 79, 85, 91, 111–113, 125
Contract price 41, 50–55, 71, 19, 82–88, 94, 99, 110

Contract, conditions of 48, 65, 66–73, 87, 89, 101, 120
Contract, engineering 6, 59, 66
Contract, reimbursable 94, 99–100, 111
Contribution 41–44, 52–56, 77–79, 82–88, 110, 123, 172
Cost center 77–81
Creditor 23, 56, 128, 167–173
Critical path 115–118, 174–177

Damages 63, 69–71
Damages, liquidated 63, 69–71, 86
Debtor 13, 127, 167–173
Delivery 99, 107–108
Depreciation 144, 167–173
Direct cost 40–43, 50, 77–81, 83–86, 100, 121, 172
Directors, Board of 27–32, 36

Environment 71, 139–140, 148–151, 152–154, 161
Ethics 148, 160–164
Extra to contract 71, 101, 109–110, 133

Force majeur 48, 69

Goodwill 20

Interest 12–16, 37–39, 52–58, 127, 167–173

Letter of credit 108
Letter of intent 64
Liability 18, 167–173
Life cycle cost 104, 138–147
Limited company, private 25–27, 37

Engineering Money: Financial Fundamentals for Engineers, by Richard Hill and George Solt
Copyright © 2010 John Wiley & Sons, Inc.